# A SOCIOLOGY OF COMMERCIAL FLIGHT CREW

*For Alex and Jack. Aviation is a fine profession. Think about it.*

# A Sociology of Commercial Flight Crew

SIMON A. BENNETT
*University of Leicester, UK*

ASHGATE

Published by
Ashgate Publishing Limited
Wey Court East
Union Road
Farnham
Surrey, GU9 7PT
England

Ashgate Publishing Company
Suite 420
101 Cherry Street
Burlington, VT 05401-4405
USA

Ashgate website: http://www.ashgate.com

**British Library Cataloguing in Publication Data**
Bennett, Simon, 1958-
A sociology of commercial flight crew
1.Air pilots - Social conditions 2.Air pilots - Social life and customs 3.Air pilots - Interviews 4.Aeronautics, Commercial - Sociological aspects
I.Title
305.9'62913092

**Library of Congress Cataloging-in-Publication Data**
Bennett, Simon, 1958-
A sociology of commercial flight crew / by Simon A. Bennett.
p. cm.
Includes bibliographical references and index.
ISBN 978-0-7546-4317-3
1. Air pilots--Great Britain--Attitudes. 2. Air pilots--Great Britain--Social conditions. 3. Airlines--Social aspects--Great Britain. 4. Aeronautics, Commercial--Great Britain--Sociological aspects. I. Title.

HD8039.A42G717 2006
305.9'629130941--dc22

2006006788

ISBN : 978 0 7546 4317 3

Reprinted 2008, 2009

Mixed Sources
Product group from well-managed forests and other controlled sources
www.fsc.org Cert no. SA-COC-1565
© 1996 Forest Stewardship Council
FSC

Printed and bound in Great Britain
by MPG Books Group, UK

# Contents

# List of Tables

# Acknowledgements

Growing up in Glamorganshire, South Wales, I developed a fascination with flying at an early age. Our house was overflown by both low and high-flying aircraft: we were overflown at low-level by Royal Air Force Hawker Hunters, at high level by passenger aircraft bound for the United States. The Hunter's noise footprint gave me just enough time to run outside to glimpse that wonderful machine as it screamed overhead (I can still call to mind the pitch of the Rolls Royce Avon's jet efflux – music to my young ears). Having got outside, I'd then watch the 707s and DC-8s heading out over the Atlantic.

After leaving Polytechnic in the early 1980s I made a big mistake ... I forgot my childhood obsession and went into financial management – lucrative, but a crushing bore. (There's an important lesson here. I should have paid closer attention to the Monty Python sketch: 'Why Accountancy Is Not Boring'.)

Eventually I came to my senses, dropped out of the 1980s' rat-race and returned to university to do a PhD in sociology. It was at this point that I decided to lead a more interesting life. I tried to get in to aviation. I succeeded. I consulted for two UK-registered carriers and became a Member of the Royal Aeronautical Society. I even trained as cabin crew.

Many people have helped me realise my boyhood dream. Here are the names of some of my helpers: Captains Peter Griffiths, Simon Searle, Ed Winter, Jim Pegram, Lance Jordan, Dave Owens, Steve Moody, John Mahon and John Moss; Go Fly's Chief Executive Officer Ms Barbara Cassani (whose hard work and can-do attitude helped secure the 2012 Olympics for Great Britain); Captain Ralph Kohn, Captain Ron Macdonald, Mr Peter Moxham, Mr Peter Richards and Mr Brian Riddle of the Royal Aeronautical Society, London; Ed Paintin of the United Kingdom Flight Safety Committee (UKFSC); Major Tony Masys of the Royal Canadian Air Force; Evan Parsons at Delta Air Lines; Mr Andy Shaw at First Choice; Ms Victoria Silverman; Dr Don Harris at Cranfield University; Professor Roger Wootton at City University; Laurie Benn at London Metropolitan University; Paul Sterbenz at North American Airlines; Simon Calder at *The Independent* (both for promoting my first book *Human Error – by Design?* and for his thoughtful writing on the aviation industry); James Eales at IFALPA; Donna Hinchcliffe, Cindy Lewis and Gemma Evans at Nottingham East Midlands Airport; Sarah Jordan; Captain Mimi Tompkins of Aloha Airlines in Hawaii (a modern heroine whose quiet strength taught me a lot about *the right stuff*).[1]

1 Mimi Tompkins was the First Officer on an aged Boeing 737 that lost part of its cabin roof due to metal fatigue. Mimi helped land the aircraft safely in a remarkable feat of airmanship. Captain Tompkins stands knee-high to a grasshopper. So much courage in such a small frame.

My sincerest thanks to the many Captains and First Officers who took the time to complete my none-too-short questionnaire and the six pilots who agreed to spend some of their valuable leisure time with me answering questions.

Thanks also to the Captains and First officers who, over the past six years, have allowed me to observe them from the jump-seat as they flew their passengers and freight to various European airports and the many, many cabin crew who have supplied me with refreshments as I made my observations from the jump-seat. (I have now brought myself to forgive the Go flight attendant who emptied a cup of coffee down the back of my neck as she entered the flight-deck ... but I still bear the scars.)

Returning to the 1970s, I would like to thank my first Professor, Bob Haigh. Bob is a remarkable man. He began his working life as a coal miner in Yorkshire. He then worked his way up through the ranks of academe to become Head of Politics at Sheffield City Polytechnic (a fine institution much missed by its alumni). I first met Bob in 1978 when I enrolled to study Public Administration (actually a thinly disguised degree in Politics). Inevitably, he made a big impression on all of us callow youths, not only because of his intelligence and kindness, but also because of his wit. I still smile when I remember how he would enter the lecture hall at nine o'clock of a December morn, stride to the windows, open them all as wide as they would go and then proceed to deliver a scintillating fifty-minute lecture on 'Lloyd George and the Decline of the Liberals' in the teeth of a howling gale. (His reasoning, of course, was that fresh air was the best cure for undergraduate hangovers.) Happy days! Thanks, Bob.

Going back a little less far (to the early 1990s) I would like to thank my Professor at Brunel University, Alan Irwin. Alan taught me how to do quality research and nurtured my interest in the sociology of science and technology. The fact that Alan supported Sunderland FC did not diminish my respect for him.

I would also like to thank my first External Examiner at the University of Leicester, Professor Brian Toft. Brian has made a significant contribution to risk management teaching at Leicester and to the discipline's development generally. His analyses of medical error should be required reading (Ministers of the Crown take note).

Finally, thanks go to my colleagues at the University of Leicester, Dr Martina McGuinness and Dr Jim Broderick. British academia is, in my experience, a vain, spiteful, dog-eat-dog world in which the term 'colleague' rarely means what it is supposed to mean (a trusted co-worker). Thankfully Tina and Jim are true colleagues. Thanks for your help. And thanks for tolerating my workaholism.

# Chapter 1

# Introduction and Basic Concepts

> As a piece of applied science the aeroplane has a place alongside the wheel, gunpowder, the printing press and the steam engine as one of the great levers of change in world history. (Rendall 1988, p. 7)

> There can be little doubt that air transport is still an extraordinarily dynamic industry. (Orlady and Orlady 1999, p. 22)

This is not a 'flashy' academic tome. It makes no great philosophic or intellectual point. But then it was never intended to. This is a workaday, user-friendly survey of the men and women who labour long and hard at the coalface of commercial aviation – the pilots. It is written in an accessible manner. Jargon – especially sociological jargon – has been kept to a minimum. Technical and sociological terms are explained. There is even a definition of sociology (see below).

The book's purpose is straightforward – to provide information to managers, regulators and other decision-makers so they can make better-informed decisions. The book complements established feedback mechanisms, like trade union representation and confidential reporting systems (for example, the UK's Confidential Human Factors Incident Reporting Programme (CHIRP)). It gives pilots another voice, and managers a further opportunity to listen and learn.

The book is based on two pieces of original research: a questionnaire survey and six in-depth interviews. Both modes of inquiry produced a significant volume of primary data (despite one interviewee's warning: 'You will be lucky to get many replies. I have distributed this questionnaire widely, but I know pilots'). The written and oral responses are reproduced *verbatim*. The data has not been 'spun' or manipulated. The intention is to paint an accurate picture of the modern pilot and her/his industry. None of the primary data has been published elsewhere. The data are classified as emanating either from a 'respondent' (a pilot who submitted a questionnaire) or 'interviewee' (one of the six pilots who were interviewed by the author).

## What is sociology?

Sociology seeks to understand the individual as a *social being*. As far as the sociologist is concerned, no woman or man is an island. To a greater or lesser extent, say sociologists, our beliefs and actions are shaped by the beliefs and actions of others

– through a process of *socialisation*. As Taylor *et al.* (1995, p. 1) put it: '[S]ociology is "the study of people in social groups" .... People are social beings .... Their lives are shared with others and shaped by the society in which they live'. Sociologists, therefore, see a relationship between an individual's beliefs, statements and actions and her/his social milieu.

To understand 'private troubles', claimed the celebrated American sociologist C. Wright Mills, one must first take account of relevant 'public issues'. Applying this maxim to commercial aviation, it follows that to understand pilots' beliefs, outlooks, actions, hopes and fears one must also understand their social history and lived experience – their class origins, educational background, current class position, family situation, social support networks, employment history, the circumstances of their current employment, access to finance, state of health, health outlook, pastimes, difficulties experienced due to gender or race, etc. In short, to get the fullest picture of the individual-as-social-actor we must take in all relevant dimensions of existence.

Sociologists treat 'common-sense' explanations of social action with determined scepticism. This scepticism, say Taylor *et al.* (1995, p. 3) '... is an aspect of the "sociological consciousness", part of what the philosopher Nietzsche called "the art of mistrust"'. The sociologist's mission is simple. It is to challenge common-sense explanations of complex phenomena with better researched, more textured, more 'scientific' accounts. As Taylor *et al.* (1995, pp. 2–4) put it: 'The sociological perspective can ... be understood in terms of such phrases as "seeing through", "looking behind", very much as such phrases would be employed in common speech – "seeing through his game" .... Just because something sounds reasonable or logical does not mean that it's true. One of sociology's most important contributions has been to challenge "obvious" statements of apparent truth'.

Social life is riddled with stereotypes. Stereotypes are '... sweeping generalisations which are highly selective and over-simplified ... stereotypes are usually factually incorrect, exaggerated or distorted ... ' (Taylor *et al.* 1995, p. 75). Harvey and MacDonald (1993, p. 37) note: 'Stereotypes are inflexible categories which are used to ascribe particular characteristics to other groups of people. The British have a stereotype of Australian men as unsophisticated, macho beer drinkers. Such stereotypes may be facile but are not necessarily dangerous. However, when stereotypes are used as a basis for prejudice ... then they are dangerous'. Where stereotypes are used to construct or frame the aviation industry, it is important that the industry challenges them with the truth. The author hopes that this book will contribute something to the aviation industry's understanding of itself, and that in doing so it will lead to a better-informed discussion about the industry's most important resource – its pilots.

Finally, a word of caution: sociology cannot generate *absolute* truth or certainty. Like any science, its methods and data are fallible (indeed fallibility lies at the very heart of natural and social science). It can, however, allow us to rise above mere prejudice in our dealings with the world around us. As Cuff *et al.* (1992, p. 9) explain: '[Sociology] has no special ability to provide the "truth" about social life .... All it can provide is a number of viewpoints or perspectives on the social world which might help us to develop greater understanding. *Occasionally, such understanding might provide a clearer basis for practical action in the world ...*' (my emphasis).

## Why a sociology of pilots?

Hudson and Pettifer (1979, pp. 6–7) note: '[T]he history of aviation is not only or chiefly about aeroplanes. It concerns people, too, the people who, in their many different jobs, have made the airlines work ... '. Rendall (1988, p. 7) makes a similar point: 'The impetus behind the scale and pace of the progress in aviation has been people – people of ingenuity, people with stark, physical courage ... '. The success of commercial aviation is due therefore to the imagination, ambition and dedication of its employees. In my opinion, commercial aviation's future prosperity depends in large part on developing more subtle understandings of those who work in the industry. Only if the industry understands its key resource can it be nurtured and used to best effect. Employees are expensive (as any wages clerk would attest). Why not make optimal use of them?

While commercial aviation has benefited from the attention of psychologists, ergonomists, management scientists, economists and other academicians, it has been more or less by-passed by sociologists – despite the fact that some occupations (coal mining and train driving,[1] for example) have profited from sociologists' input. Of course there may be sound practical reasons for this. In today's strained security environment gaining access to the 'sharp end' of commercial aviation is very difficult. It was difficult after the terrorist attacks of September 11th, 2001. One can be sure that it will be even more difficult in the aftermath of the July 2005 attacks in London. More to the point, perhaps, the highly competitive nature of aviation *the enterprise* tends to work against an 'open door' policy for academics – however noble their intentions.

## Research method

This book is an ethnographic study of pilots. An ethnography is a detailed account of the origins, social complexion, motivations, behaviour and aspirations of a particular social group – in this case commercial pilots. Much of the account consists of subjects' personal statements, which are reproduced *verbatim*. These statements, some of which are quite long, constitute a 'popular history' of the industry. In the context of this research, Gilbert's (1993, p. 157) definition of ethnography is apposite: '[T]he techniques are likely to include interviews (usually more like a conversation than a standardised interview ...), the analysis of documents, direct observation of events, and some effort to "think" oneself into the perspective of the members, the introspective, empathetic process Weber called "verstehen"'. While this particular

1 With regard to sociological studies of train drivers, see, for example Heath *et al.*'s (1999) study of London Underground tube train drivers.

study lacks the element of direct observation, the author has been able to draw upon his previous longitudinal ethnographic studies of commercial pilots (stretching back over a period of some six years).[2] The many hours spent in crew rooms, on the ramp and on the flight deck by the author go some way to satisfying Weber's 'verstehen' criteria.

As mentioned above, one of the objectives in writing this book was to give commercial pilots a voice. In providing this voice, the book supplements such anonymised reporting mechanisms as the Confidential Human Factors Incident Reporting Programme (CHIRP) and the PPRrune website (CHIRP and PPRrune are UK-based independent reporting facilities). The book's method is informed by the venerable ethnographic tradition of allowing subjects to 'speak for themselves' (this is sometimes referred to as oral history). This facilitative technique has been used to good effect in numerous social and industrial contexts. There is, for example, the Greater Manchester Low Pay Unit's (GMLPU's) study *Workers' Voices: Accounts of Working Life in Britain in the Nineties*. *Workers' Voices* is described as: '[A]n edited transcript of interviews undertaken with twelve people ... ' (GMLPU 1995, p. ii). There is also the Centerprise Trust's *A People's Autobiography of Hackney*. Published in 1977, this oral history of the London Borough of Hackney is based on in-depth interviews with 13 Borough residents. Finally, there is the definitive Mass-Observation project initiated in Britain before the Second World War.

## Mass-Observation – the definitive ethnography

The organisation known as Mass-Observation was founded in 1937 by three social scientists who '... felt that there was a gap in real knowledge about the lives of ordinary people'. Mass-Observation's founders set out to develop a more textured account of life in Britain than that offered by, for example, the mass-media. According to Bolton Museums (2005) the British media: '[G]enerally portrayed the population as having a broad consensus about the issues of the day'. Mass-Observation's diarists and correspondents produced a detailed ethnographic account of the lives of ordinary Britons: 'A team of paid investigators went into a variety of public situations: meetings, religious occasions, sporting and leisure activities [the pub, for example], in the street and at work, and recorded people's behaviour and conversation in as much detail as possible .... [The] aim was to create an "anthropology of ourselves"' (Mass-Observation Archive 2001).

Very much in the tradition of Mass-Observation, *A People's Autobiography of Hackney* paints a picture of life in the Borough in the late 1970s by giving those who knew most about it – Hackney residents – a voice. Centerprise researchers conducted interviews with a barber, hairdresser, mortician, health visitor, postman, shop worker, housewife, taxi driver, fish and chip man (a man who sold fish and chips in a shop), teacher, machinist (a woman who sewed dresses), curate and school-

2 See, for example, Bennett (2003) 'Flight crew stress and fatigue in low-cost commercial air operations – an appraisal'.

leaver. This generated a remarkably insightful and textured account of the 'lived experience' of Hackney residents. Fifteen years ago *A People's Autobiography of Hackney* inspired this author to use the oral history tradition to inform his academic method. If *A Sociology of Commercial Flight Crew* is even half as insightful as *A People's Autobiography of Hackney*, then the three years spent researching and writing it will not have been wasted.

## The researcher

All the primary data used in *A Sociology of Commercial Flight Crew* were gathered by the book's author. The author's familiarity with the commercial aviation industry informed both the choice of research method and subsequent analysis and presentation of data. The author has worked closely with UK flight and cabin crews for over six years. He has worked alongside them, talked with them, interviewed them[3] and socialised with them. He has also trained with them. Before flying jump-seat, the author was required to pass the British-Airways designed and delivered Safety Equipment and Procedures (SEP) course. The author has flown jump-seat on both Boeing and Airbus aircraft and has supervised passengers on the ramp during turnaround. He has also written about the safety of ramp operations.[4]

## Structure of the book

Despite the book's emphasis on original research and primary data, it was felt that an interpretative framework should be provided to aid comprehension. To this end, Chapter 2 – Pilots' History provides a social, economic and political context for pilots' historic experience of commercial aviation. Through the medium of *secondary* data it reviews pilots' experience of aviation from the early days of mail planes and flying boats to the far-from-glamorous world of regional and low-cost carriers – the bread-and-butter operations of today's industry.

References to wider social, economic and political dynamics (like the post-Second World War consensus, the two 'oil shocks' of the 1970s, Reaganism, Thatcherism, Blairism (that is, 'Third Way' politics), monetarism, globalisation and the rise of the Chinese and Indian economies) are used to build perspective. This emphasis on context reflects the author's view that to understand the particular (in this case pilots' historic experience of commercial aviation) one must first understand the general (the aviation industry's social, economic and political context).

Inevitably, any industry is a product of its time. An industry's structure and mode of operation will reflect the values of the society from which it has sprung. Take

---

3 The journal paper 'Flight crew stress and fatigue in low-cost commercial air operations – an appraisal', for example, reproduces extended interviews with Captains and First Officers.

4 See Bennett and Shaw (2003).

the British experience, for example: in the 1940s, 50s and 60s the commitment of successive UK governments to the mixed economy and post-war settlement[5] created an environment that favoured state-subsidised flag carriers such as BEA and BOAC. BEA and BOAC were *political instruments*. The later disintegration of the post-war settlement and consequent ideological renaissance of laissez-faire economics signalled the end of the airline-as-political-instrument model. Thatcherism (an Anglicised Reaganomics) changed the complexion of the industry. Market entry was encouraged through an emphasis on enterprise. British Airways was privatised.

The British experience demonstrates the aviation industry's symbiotic relationship with politics and broader social forces (like consumerism and liberalism). The industry cannot be understood in isolation. It is a social artefact – a product of humankind's ambition and imagination and politicians' informed choices ... or dogma.

The questionnaire survey and interviews generated a significant volume of data, much of it in narrative form. Rather than require the reader to digest every word of testimony, pilots' responses are reproduced verbatim in the several Appendices. (The use of Appendices to separate volumes of data from concise analysis has been used to effect in, for example, Kohn's (2005) *So You Want To Be A Pilot 2006* and Wood's (2004) *CAA Paper 2004/10. Flight Crew Reliance on Automation.*) The responses are summarised and analysed in five themed chapters, specifically: 'Chapter 3 – Pilots' Socio-economic Origins and Motivations'; 'Chapter 4 – Pilots' Lifestyle and Socio-economic Status'; 'Chapter 5 – Pilots' Commercial Flying Careers and Perceptions of the Industry'; 'Chapter 6 – Pilots' Concerns and Professional Culture'; and 'Chapter 7 – Pilots' Aspirations and Future Plans'. 'Chapter 8 – Conclusions' draws together the several themed analyses to paint a picture of the pilot as a social actor. That is, someone with a history; a work life; a home life; a social life; someone caught up in numerous social and economic networks. In other words, a human being ... hopeful and fallible. The final chapter is the most ambitious. 'Chapter 9 – An Agenda for Managers' attempts to synthesise pilots' (sometimes contradictory) opinions and beliefs into an 'action list' for decision-makers (airline managers, regulators, politicians, researchers, etc). Its purpose is to raise decision-makers' awareness of pilots' ideas and concerns.

## Methodological weaknesses and omissions: *mea culpa*

In a spirit of reflexivity[6] it is acknowledged that the book's methodology has certain weaknesses. For example, while the questionnaire survey was widely advertised, the

5 The British post-war settlement involved an acceptance by both the Labour and Conservative parties of the government's role in maintaining full employment and economic growth. There were numerous sub-settlements: industrial; economic; social; gender; racial; environmental; regional; national, etc. (Hall and Jacques 1989, pp. 25–29).

6 In the context of academic research, being reflexive means making explicit the practical and theoretical limitations/constraints of one's research method(s). This allows the reader to judge the integrity and significance of the data generated by the research.

response rate did not run to the thousands (although it was heartening that responses were received from around the world). Also, time constraints provided for only six in-depth interviews to be conducted (although the use of a semi-structured interview format allowed interviewees to expand on topics addressed by the interviewer and explore new ones. Some of the interviews ran for several hours. All the interviews were taped, then transcribed *verbatim*. This ensured that pilots' statements were accurately recorded).

Due partly to the author's provenance and the fact that the research was (perforce) conducted from the UK, it is the British and American experience of aviation that looms largest in Chapter 2 – Pilots' History and at various other junctures in the book. This Anglo-American axis is not intended as a sleight on the rest of the world, however. It is recognised that many countries have made valuable contributions to aviation. One need only think, for example, of the Australian aviation community's contribution to safety assurance, the remarkable success of Brazil's youthful aircraft industry,[7] Switzerland's contribution to short take-off and landing (STOL) technologies, and Canada and the Soviet Union's achievement in using their own rugged designs to exploit their natural wealth, to understand that the industry is the product of a gargantuan *global* effort.

Finally, certain issues were not addressed in either the interviews or via the questionnaire. One such issue is the impact of seniority lists[8] on pilots' careers and levels of job satisfaction. This omission was highlighted by a respondent, who wrote: 'The questionnaire did not address the impact of seniority lists on pilots' careers, both in terms of job security and safety'. Clearly this issue merits detailed investigation. The author can only apologise in advance for this and other omissions.

## How to read this book quickly

This book contains a large volume of primary data and comment. Some may not wish to negotiate every sentence of the book's data, analysis and argument. For those with little time to spare, most Chapters end with summaries. The addition of

---

7 Theobald (1994, p. 25) notes: 'Some impression of the extent to which economies such as that of Brazil have developed over the past twenty years may be gathered from the fact that in the late 1980s the Royal Air Force ordered 130 wholly-Brazilian-engineered training aircraft'. The Embraer Tucano is the RAF's basic trainer. Made under licence by Shorts of Belfast, the turboprop-powered Tucano S312 has a top speed of about 270 mph (Rendall 1996, p. 199).

8 The more traditional 'legacy' carriers generally operate a seniority list system of promotion, whereby those who have been with the company the longest are considered first for promotion. To this end new recruits are given a number when they join the company (this number is often the date on which they joined). Some new airlines have dropped the list system in favour of promotion-on-merit. Given that both systems have major implications for pilots' career prospects (and life chances) they merit close examination. It might be found that efforts to remove the list system reflect wider efforts to establish more meritocratic societies.

summaries reflects the author's desire to make the book as accessible as possible. Having said this, there is much to be gained from perusing *A Sociology of Commercial Flight Crew*'s primary data, almost all of which is reproduced in the Appendices. Unexpurgated work is invariably more interesting, if a little more demanding.

# Chapter 2
# Pilots' History

## A panoptic

> The airman ... delights in the ever changing patchwork of light and shadow on the earth's surface, the toylike aspect of everything beneath, the wonderful variety of patterns which roads, rivers, valleys, forests and hedgerows provide in unbroken succession. (Supf cited in Duke and Lanchbery 1964, p. 243)

> A Boeing 707 captain, looking back at the huge aircraft he [sic] has just brought down through fog to a silk-soft landing, has every right to be proud of the disciplined skills which enabled him to do it .... If a magic sort of X-ray photograph that showed the psychological factors inside him could ... be taken ... they would reveal an intelligent human being, conscious of his enormous responsibilities ..., highly skilled, courageous, conscientious – and very vulnerable. (Beaty 1969, pp. 16–25)

> Are we going to apologise when something goes wrong? No, we're f***ing not. Please understand. It does not matter how many times you write to us complaining that we wouldn't put you up in a hotel because there was fog in Stansted. You didn't pay us for it. (O'Leary cited in Johnson, 2004)[1]

> We should shoot EU regulators and the airlines might be able to prosper. (O'Leary cited in Osborne, 2005)

Commercial aviation is constantly evolving. The above quotations capture the zeitgeist of the three ages of commercial flight, from the revelation, romance and risk of aviation's pioneering days, to the technological advancement and studied seriousness of the Jet Age, to the unbridled commercialism and boldness (if not crassness?) of the industry in the New Millennium.

Pilots, of course, have always been at the heart of commercial aviation. Without them there would be no industry. Technology may be able to navigate, trim and land aircraft automatically, but, for the time being at least, no-one is seriously suggesting that the pilot be removed from the flight deck (although the 1980s did see the flight engineer substituted by automatic devices like the flight management computer (FMC)).

---

1 In 2004 Michael O'Leary was the Chief Executive Officer (CEO) of Irish-registered low-cost carrier Ryanair.

Today, the pilot's experience of flying is, to varying degrees, mediated by technology and the cocooning effect of the air-conditioned, sound-proofed and ergonomically-correct flight-deck. In the early days the pilot's experience of flying was direct and immediate. Her/his relationship with the aircraft and its operating environment had a physical, visceral quality. The stick was connected to the aircraft's various control surfaces by wires. The control system was as direct and intimate as it could possibly be. Early flying was 'seat-of-the-pants' flying. It was intuitive. Pilots were admired for their stick-and-rudder skills. This was the Age of Airmanship – of *airmen*. Pilots could feel the aircraft vibrate and shudder. They could hear it creak and groan. They could feel the wind and rain against their faces. They could smell the engine exhaust and oil. For those who would describe this period as flying's 'romantic age', K.C. Gandar Dower's (1934) account of his flight to Dijon in a Puss Moth is required reading: 'The flight to Dijon was thoroughly unpleasant. In flying, whether you like it or not, you are mentally, morally, and physically the slave of the elements. On the ground it can only rain on your head; in the air you are completely surrounded by weather ... [W]hen the weather is foul there is no peace or poetry at all. Peace and poetry have never been within 100 miles of having anything to do with this kicking, jerking, roaring, vibrating, plunging, abominable body of sickness .... Great up-currents hit you from below, and you soar dizzily. Then the seat falls away; there is a second in which you are literally in suspense. Everything comes down on everything with an immense bump, and you feel the shudder of the strain upon the wings .... Whatever does anyone want to fly for? An expensive and uncomfortable method of suicide. A kind of switchback railway extended – oh, extended very much too far!' (Gandar Dower cited in Duke and Lanchbery 1964, pp. 246–247).

On the other side of the Atlantic, airmen were having identical experiences, as may be gathered from this testimony from pioneer flyer Jimmy Doolittle (cited in Stone and Babcock 1988, p. 531): '[A]bout four hours out, I ran into solid overcast and severe thunderstorms .... The air was extremely turbulent and the airplane was violently thrown around .... Although the DH-4 had excellent stability characteristics, I could hold it on a relatively even keel only with great concentration and effort. After the lightning died away, the turbulence seemed to intensify and there was about an hour in the jet black darkness when no ground reference point could be seen .... Although I had been flying for almost five years "by the seat of my pants" and I considered that I'd achieved some skill at it, this particular flight made me a firm believer in proper instrumentation for bad weather flying'.

According to Stone and Babcock (1988, p. 530) the early pilots were intimately involved not only with their aircraft but also with the designers and manufacturers of those aircraft: 'In those early days of aviation there was a direct interaction between the pilot, the builder, and the aircraft engineer. Often the pilot was included from the first day of design concept to the roll-out and initial flight of each aircraft .... [P]ioneers in the field of all-weather flying [like Doolittle] had great influence in the initial design of cockpit instrumentation layout'.

If human beings are as much a product of nurture as nature, as much a product of experience as innate character, then episodes like the ones described by K.C. Gandar

Dower and Jimmy Doolittle would have honed certain useful capacities, like the ability to think through and solve problems *in situ* with limited resources, absorb mental pressure or endure physical discomfort. Many years after the pioneering flights of Gandar Dower and Doolittle, American popular author Tom Wolfe tried to identify the qualities that make a good test pilot. Wolfe used the term 'the right stuff' to describe these qualities. A pilot with *the right stuff* never gives up. Clever, skilled, resourceful and physically and emotionally resilient, when in difficulty s/he coolly and methodically applies potential solutions until s/he finds one that works ... or rides the aircraft into the earth or ocean in the attempt. Wolfe (1991, pp. 23–24) writes: 'A young man might go into military flight training believing that he was entering some sort of technical school in which he was simply going to acquire a certain set of skills. Instead, he found himself all at once enclosed in a fraternity. And in this fraternity ... the world was divided into those who had it and those who did not. This quality, this *it*, was never named .... As to just what this ineffable quality was ... well, it obviously involved bravery. But it was not bravery in the simple sense of being able to risk your life ... any fool could do that .... No, the idea here ... seemed to be that a man should have the ability to go up in a hurtling piece of machinery and put his hide on the line and then have the moxie, the reflexes, the experience, the coolness, to pull it back in the last yawning moment – and then go up again *the next day*, and the next day ... '.

Wolfe is not the only author to have attempted a definition of *the right stuff*. In his book *Airline: The Inside Story of British Airways*, Arthur Reed (1990, p. 110) quotes from a schoolchild's essay titled 'What I Want to Be when I Grow Up': 'Pilots should be brave so they won't get scared if it's foggy and they can't see, or if a wing or a motor falls off, they should stay calm so they'll know what to do'.[2] Evidently the question of what makes a good pilot is pondered by young and old. Pilots, too, wonder about *the right stuff*, as evidenced in this 32 year old female First Officer's testimony: 'An aptitude for flying is an obvious must. This is a combination of hand-eye coordination and the ability to do very simple mathematics at the same time. Spatial awareness and a scientific tendency are also useful. Communication is another essential skill. Modern airline companies use lots of computers, but rely heavily on the teamwork of humans on the flight deck and in the aircraft cabin. It helps to be relaxed, especially as things beyond your control, like lost passengers, luggage and airport delays, can create high stress levels .... I try and stay calm, whatever the day throws at me. Getting annoyed and shouting at someone, especially if English isn't their first language, doesn't really get you anywhere' (McLaurin 2005). Kohn (2005, p. 27) offers this summary of the personal qualities needed by pilots: 'Enthusiasm and motivation for flying need to be complemented by a self confident, calm, reliable, disciplined, level-headed and responsible attitude'.

It is not uncommon for contemporary flight crew and their representative organisations to complain about what they perceive to be 'excessive' hours. In 2002 a former pilot at an Irish-registered low-cost carrier (LCC) told *The Observer*

2 Published in British Airways' Boeing 757 Flight Crew Newsletter.

newspaper: 'The culture at that company is one of pressuring pilots to work very long hours and it has only got worse since I left' (cited in Walters, 2002). The Assistant General Secretary of Impact, the Irish union that represents air traffic controllers, pilots and other aviation industry staff remarked: 'There is a very strong, unwritten culture that you do not question the way the company is run' (cited in Walters, 2002). According to Walters (2002) the LCC's Chief Executive Officer (CEO) '... fiercely rebutted all the allegations and said his pilots "did a fantastic job", never flew "over hours" and were encouraged to raise safety issues with senior staff'.

Pilot–management tensions are far from new. As Hudson and Pettifer (1979, p. 104) note, even in aviation's romantic pre-Second World War era pilots found themselves the subject of exploitative terms and conditions: 'How ... did the airlines ... manage to stay alive? Wages and salaries, of course, were low ... strikes and "industrial action" were rare. The companies ... managed to get a prodigious amount of work out of their pilots'. Speaking about his terms and conditions at Imperial Airways (one of British Airways' precursor companies) one pilot remarked: 'We were paid on a somewhat iniquitous system of a low basic salary and so much a flying hour. Greed therefore forced me to fly as many hours as I could. I used to try to average about a hundred hours a month ... ' (cited in Hudson and Pettifer 1979, p. 104).

It is noteworthy that this pilot's personal flying hours target was the same as the legal maximum in force at the time of the 2002 *Observer* article (that is, 'Not more than 100 hours total flight time per 28 rolling days' (International Transport Workers' Federation, 2004)). It would be interesting to know what this Imperial Airways pilot would have made of the salary levels that obtained in 2002. Two years into the New Millennium jet pilots could earn between £50,000 and £125,000 (Walters, 2002). Interestingly some of today's pilots are paid on the same basis as those who flew before the Second World War – they receive a basic salary that they augment with sector pay and other output-related bonuses.

Hudson (1972, p. 19) suggests that the 1920s pilot had two incentives to getting aloft: first, the need to augment her/his basic salary; secondly, the desire to be seen as someone who could get the job done come what may: '[S]ome part of a pilot's reputation was made by his ability to take off and land under conditions which grounded less skilful and less daring men. There was certainly a financial incentive to fly whenever possible, because a pilot received extra money when he was actually in the air. Shaw [Captain Shaw, the pilot who flew the first international scheduled air service in 1919] had £700 a year while he was with Air Transport and Travel, together with 50p (10*s*.) an hour flying pay'.

While today's turbofan or turboprop-powered aircraft are much more complex than the piston-engined aircraft of the 1930s and have to navigate significantly more crowded airspace to land at significantly busier airports, they do at least have adequate instrumentation, the option of using autopilots, flight management computers, weather radar, terrain radar, ground-proximity warning systems (GPWS), traffic alert and collision-avoidance systems (TCAS) and other technical aids. They also have a relatively comfortable working environment. The pilots of yore had to fly accurately and safely in aircraft whose wings were made of canvas stretched over a wooden frame and whose cockpits were devoid of all but the most basic of instruments and home comforts. As Robert Buck, a pioneer aviator who ended his career flying Boeing 747s, explains: '[I]n 1931 ... I was in a Pitcairn Mailwing – an open biplane,[3] 225 horsepower, no radio or fancy instruments, only a simple giro instrument called a turn-and-bank indicator, which had a ... vertical hand that swung left or right with the airplane's nose, and a curved carpenter's level with a steel ball instead of a bubble to show slip and skid. Airspeed and rate of climb instruments ... made up the complement of blind-flying instruments' (Buck 1994, p. 6).

While airways and airports were less crowded in Buck's day, point-to-point navigation could involve flying at low altitude to follow a railway track, canal, river or road.[4] Early aviation, like today's low-cost sector, had a back-to-basics feel about it. As Hudson and Pettifer (1979, p. 104) explain: 'Those were the days when pilots bicycled to the airport and sustained themselves on sandwiches and a flask of coffee brought from home. The salaries they were paid demanded simple tastes'. However, while life was hard for the pioneers of commercial air services, there were compensations, like the possibility of achieving an ego-boosting (although relatively minor) degree of celebrity: '[A] number of [pilots] became well-known public figures. Several airlines, among them KLM and Imperial Airways, issued passengers with signed, postcard-sized pictures of the pilot in charge of their flight. Regular travellers built up a collection of these cards' (Hudson and Pettifer 1979, p. 104).

While pilots were required to endure spartan conditions, their charges were provided with increasing levels of comfort. In the late 1920s, Imperial Airways introduced their *Silver Wing* service between London and Paris using the Armstrong-Whitworth Argosy, a 20-seat, three-engined biplane. Taylor (1967, p. 76) offers the following description of this pioneering service: 'New standards of luxury were introduced when Imperial Airways opened their *Silver Wing* lunchtime service between London and Paris in 1927. The two rear seats ... were replaced by a buffet and steward's position ... [but] the pilots still sat side-by-side in an open cockpit'. Photographs of the cabin show an interior with electric lighting, curtained windows, comfortable-looking seats with padded head-rests and a drinks station presided over by an immaculate, white-jacketed Steward.

It was perhaps fortunate that many of those who piloted such early, open-cockpit types as the Argosy were hardened ex-combat flyers from the Great War. As Rendall (1988, p. 115) explains: 'Britain had trained 26,000 pilots: most were young men who went straight from school to war with little experience of life in between.

---

3 Early passenger aircraft, too, had open cockpits. In aircraft like the Boeing 40B and de Havilland D.H.66 Hercules, the pilot sat in an elevated open cockpit. Today there are complaints that locked cockpit doors and a ban on passenger visits to the flight deck have served to alienate flight crew from cabin crew and passengers (Bennett, 2002a). But of course the same situation obtained in the early days of flying with aircraft like the Boeing 40B and D.H.66. The pilot(s) braved the elements while their passengers sat in relative comfort. Some aircraft had stewards who served passengers with food and drinks.

4 In the USA this form of navigation was known as 'contact flying'.

Plucked from a gruelling yet stimulating experience they were restless, press-on types looking for exciting outlets for newly redundant skills'.[5] It was not until the introduction of aircraft like the Fokker F.VIII, Handley Page H.P.42 and Ford Tri-motor that pilots were given enclosed cockpits.

It can be seen that flying commercial aircraft in the first half of the Twentieth Century was a physically demanding and not very well paid occupation. There were few comforts and plenty of hazards. In the 1920s, the dearth of on-board flying aids meant that pilots had to rely on whatever *intuitive* skills they possessed. According to Beaty (1969, p. 2) pilots' sense of self-worth correlated with their capacity for intuitive flying. The better they were at seat-of-the-pants flying the better they felt about themselves: '[I]n the 1920s ... [p]ilots expected to be able to fly in cloud and at night without instruments. It was considered that this was the sort of skill that a *real* flyer possessed, and you were rather less of a man and pilot if you couldn't do it'.

The early pioneers were obliged to live with risks that would be considered unacceptable by most of today's commercial pilots (although, if Tom Wolfe is correct, not by test pilots). In his history of the development of air traffic control, *Take-off to Touchdown*, Charlwood (1968, p. 2) comments: '[L]ike all pioneers, pilots had to take risks. Through great skill most of them survived. But there were others ... who eventually met a day when circumstances were all against them. Then their flights ended in disaster. From the study of such disasters governments slowly learned the need for aids to navigation ... '.[6]

If culture is 'the way we do things around here', then the culture of the pioneer aviators of the first half of the Twentieth Century was one of risk-acceptance and, on occasion, risk cultivation. Commercial flying had a macho quality. It was (perforce) as much a test of emotional and physical resilience as of technical competence. Not unsurprisingly, airline managements sought to cash in on their employees' daring exploits and devotion to duty. As Hudson and Pettifer (1979, pp. 155–156) explain: 'From the beginning the airlines had devoted a lot of time, money and effort to telling

---

5 Wars have a habit of underwriting peace-time innovation and growth. Reed (2002, p. 3), claims that Cold Wars have the same effect). Both the Great War and Second World War provided large numbers of pilots and aircraft for fledgling airlines. Orlady and Orlady (1999, p. 21) note: 'After the [end of World War II] pilots returned to their civilian jobs. There was a substantial supply of trained pilots available for expansion'. Their character and skills forged in the heat of battle, wartime pilots had sufficient of *the right stuff* to overcome the many technical and other challenges they were to encounter as pioneers of commercial aviation. Those who flew in the Berlin Airlift drew heavily on their wartime experience.

6 Charlwood makes an important observation. In commercial aviation, as in other risk-laden activities like motor sport, it often takes a disaster to persuade the industry to address a particular latent error (or 'accident waiting to happen'). The aviation industry is littered with examples of tombstone regulation and tombstone engineering (post-hoc regulatory and technological fixes). Today's traffic alert and collision avoidance system (TCAS), for example, had its origins in mounting concern about mid-air collisions in the US in the 1950s and 1960s. As the body-count escalated, so did the clamour for action. Politicians, regulators, manufacturers and operators had their hand forced.

the public what selfless, responsible men pilots were .... One interesting thread that runs through the history of passenger flying is the nature of its hero-worship at any given time. Throughout the 1920s and 1930s ... there is no doubt that pilots had almost god-like status, both for what they did and for what they were'.[7]

Eventually progress, in the form of 'the invisible hand of the market', national and international regulation and technological advance caught up with the pioneer aviator and his nascent industry. In the United States, for example, there were several important – if not revolutionary – developments. The USA's fecund domestic market spawned that country's 'big four' airlines – American, Eastern, United and Trans World (Morrison and Winston 1995, p. 3). The USA gained a de-facto 'flag carrier', too, in the form of the adventurous and technologically innovative Pan American World Airways.

For pilots, Pan American was a glamorous (and relatively safe) proposition: 'The view in Washington in the late 1920s was that the US needed a foreign airline to compete with the Europeans and to protect its investments and influence .... The US needed its "flag carrier" .... Pan American ... fulfilled the "imperial role" of airlines far better than any of its European counterparts .... Technically it was far ahead too, pioneering radio navigation .... Its aircrews in their smart, military-style uniforms became ambassadors, exuding the confidence that America was the world's leading country in commercial air transport. By 1939 it unquestionably was. As a flag-carrier Pan American had no equal' (Rendall 1988, pp. 148–153).

In 1934 the National Labour Board (NLB) sought to limit pilots' flight duty time. Stone and Babcock (1988, p. 532) explain: 'In 1934 the ... NLB ... issued what has come to be known as "Decision 83", which limited pilots at five airlines involved in a dispute (American, Northwest, Trans World, Pan American and United) to a monthly maximum of 85 flight hours (flight hours begin at engine start and taxi, and end at shutdown) .... The decision of the NLB was based on expert medical testimony of the Aeromedical Association .... While there appeared to be a difference between "Decision 83" and the final rules issued by the [Federal Aviation Administration], the NLB decision had the force of federal law ... '. President Roosevelt was determined that commercial aviation should put safety at the top of its agenda, saying on one occasion: 'Public safety calls for pilots of high character and great skill .... Therefore the law should provide for a method to fix maximum hours and minimum pay' (Roosevelt cited in Petzinger 1995, p. 220). In 1938, US legislators passed the Civil

---

7 Flying during the 1920s and 1930s was, for the most part, a male occupation. As Petzinger (1995, p. 219) notes, females were not made to feel particularly welcome: 'To the extent that flying was deemed a masculine endeavour, it was all but forbidden for women. The writer Antoine de Saint-Exupéry [an early aviator] once declared: "Flying is a man's job and its worries are a man's worries"'. Petzinger (1995, p. 219) also notes: 'A 1920s-era ad for a flying school appeared under the headline "The Aviator – The Superman of Now", to which it added, "Flying is the greatest sport of red-blooded, virile manhood". The official history of the pilots' union notes that men had always been drawn to piloting in part for "the looks they got from attractive young women"'.

Aeronautics Act which, according to Morrison and Winston (1995, p. 3), '... was to regulate air fares and routes with a heavy hand for the next forty years'.[8]

For a country committed to such libertarian ideals and doctrines as 'life, liberty and the pursuit of happiness' and free and fair competition (as opposed to collusion and market-fixing), the federal government in the 1930s took a remarkably close interest in the aviation industry in general, and pilots in particular. It could be argued that in regulating pilots' working conditions and subsidising Pan American the federal government was as interventionist as any liberal democracy of the day (although these policies are perhaps less surprising when viewed against the terms of Roosevelt's New Deal).[9]

Regarding technological advance, companies like Lockheed, Boeing and Douglas began to produce all-metal, multi-engined monoplanes with autopilots, control surface trim-tabs (the use of which by the pilot provided for a more balanced aircraft that was easier to fly), navigation aids, reliable communications, fully-enclosed and relatively comfortable flight decks, de-icing equipment, good handling characteristics and (for the day) sprightly performance.[10]

During the first half of the Twentieth Century, there was a gathering realisation that international air transport had to be regulated by legal instrument. Milde (1977, pp. 444–445) writes: 'The emergence and development of aviation ... created complex social relationships among individuals, corporations, governmental authorities and between States. These social relationships require[d] legal regulation .... The international character of civil aviation necessitat[ed] a high degree of uniformity of legal regulations'. There were many urgent needs to be met. These included the

8 According to Morrison and Winston (1995, p. 4) the dynamics of the US industry remained largely unchanged until the passing of the 1978 Airline Deregulation Act (ADA), a determinedly consumerist piece of legislation. Sampson (1984, p. 136) notes: 'President Carter, to many people's surprise ... became the ultimate deregulator [setting in motion] the greatest upheaval in American air transport since the thirties'. Carter's ADA set in motion a *global* airline deregulation movement (boosted, of course, by economic globalisation) that today (2006) is changing even India's home market, for so long hamstrung by nationalist ideology.

9 The various New Deal interventions in the US economy were intended to drag it out of its post-Wall Street Crash depression. Maidment and McGrew (1991, p. 26) comment: 'Roosevelt's New Deal ... fostered the belief that the federal government had both the responsibility and the capacity to deal with the nation's social and economic ailments'. It is interesting to contrast Roosevelt's taste for interventionism with Carter's championing of a free market in US domestic aviation in the late 1970s (that led to the enactment of the Airline Deregulation Act).

10 The Boeing model 247 had all these features. Taylor (1967, p. 87) comments: 'First twin-engined monoplane able to climb on one engine with a full load, the 247 revolutionised air transport ... '. In the same year as the 247 made its maiden flight (1933) Imperial Airways was operating the Handley Page H.P.42 four-engined biplane. According to Taylor (1967, p. 77): 'Anthony Fokker once described the Handley Page H.P.42 airliners ... as having built-in headwinds'. While they could manage only 100 miles per hour in the cruise, they did at least have enclosed flight decks.

need to establish a licensing system for flight crew, the need for a *lingua franca* for communication, the need to strike a reasonable balance between economic efficiency and passenger safety, and the need to resolve issues of liability and insurance. Some sought answers in bilateral agreements ('bilaterals'). Others looked to the establishment of independent, standard-setting supra-national bodies.

The perceived need for the management of international operations according to agreed standards was a manifestation of that most pervasive of Twentieth Century ideologies – scientific rationality. Scientific rationality promoted – and was promoted by – *industrialism*. Theobald (1994, p. 63) explains: 'The keynote of ... Industrial Society is scientific rationality, that is to say, the application of scientific thinking not only to production ... but to society in general. The emphasis is ... upon planning, the planning of production ... communications ... of social relations generally .... [W]hilst tensions and conflict are bound to develop it is possible to establish institutional arrangements which are capable of resolving them .... [M]echanisms of conciliation ensure that grievances can be aired and differences resolved [today the International Civil Aviation Organisation (ICAO) constitutes one of commercial aviation's "mechanisms of conciliation"]'.

In its purposefulness, the physical capital of commercial aviation (the aircraft, radio masts, radio beacons, airports and aerodromes) both reflected and reproduced the ideology of scientific rationality. Scientific rationality is intolerant of frivolity or indulgence. The design processes that produced such aircraft as the Boeing 247, Douglas DC-2 and Lockheed Constellation were dominated by considerations of economy, safety and performance. Commercial aircraft placed certain *invariant* demands upon their owners and operators (pilots, maintenance engineers and ground-handlers). Aircraft required and reproduced rational behaviour. They were (and are) globetrotting advertisements for, and advocates of rationalism (see, for example, Bennett 2005).

Although the first law pertaining to aircraft was passed in France in 1784, the first international convention was agreed in 1919. The International Convention on Air Navigation provided for the establishment of the International Commission for Aerial Navigation. This was followed by other international conventions, such as the Pan-American Convention of 1929 that '... included detailed technical legislation as well as some rules of private air law' (Milde 1977, p. 444). The Convention's rules were applicable only to its 16 contracting states. It was not until the 1940s that a body with a wide membership and truly comprehensive remit was established.

In 1947, the United Nations established the International Civil Aviation Organisation (ICAO), the purpose of which was to '... develop the principles and techniques of international air navigation, and to foster the planning and development of international air transport to insure the safe and orderly growth of international civil aviation ... ' (Milde 1977, p. 445). The other major regulatory innovation was the inception of the International Air Transport Association (IATA) in 1945. IATA, in effect a corporation, concerned itself with the setting of air fares, design of documentation and various other administrative necessities of international air travel.

By the middle of the Twentieth Century therefore, commercial aviation was no longer an ideologically-fuelled free-for-all scrap. Certainly there was room for enterprise, but with the advent of flag carriers able to draw on the national exchequer and the regulation of air fares by IATA (according to Milde, (1977, p. 445), a cartel), *truly* free enterprise had less room for manoeuvre than it did in the 1920s and 1930s. Nation states began to establish their own regulatory agencies. In the USA, for example, Roosevelt's interventionist instincts and distrust of the US 'majors' prompted him to establish the Civil Aeronautics Authority.[11] As Sampson (1984, pp. 46–47) explains: 'Roosevelt ... created a new Civil Aeronautics Authority to allocate and supervise air-routes – later re-formed in 1940 as the Civil Aeronautics Board. The new board perpetuated the system of "controlled competition" by which the airlines, *behind an appearance of total free enterprise*, depended for their routes on government approval – and ultimately on the approval of the President who could and often did overturn his board's judgement. *The air was becoming still less associated with freedom of enterprise*' (my emphases).

According to Petzinger (1995, p. 14) the CAB was nothing less than 'an old-fashioned statist controller' that provided for 'the involvement of the federal government in every financial aspect of the industry's business ... '. Shearman (1992, p. 85) comments: 'Initially the industry was seen to be strategically important and an infant industry which therefore had to be nurtured. Economic regulations were imposed through the [CAB] which awarded routes to airlines on application but according to the level of traffic on the route. Thin routes, i.e. those with little traffic, could be operated as a monopoly ... while major routes could support up to three carriers. Prices were fixed by the CAB on the basis of the costs of an efficient operator on each specific route or group of routes. Large quantities of cost data had to be submitted to the CAB by all airlines'. The federal government's regulatory exuberance impacted the pilot community, too. An ever-increasing number of rules, regulations and procedures meant that pilots themselves had less room for manoeuvre – sometimes literally, as with the USA's introduction in the mid-1940s of air corridors for the management of *en route* traffic. Air corridors were subsequently introduced over the North Atlantic for the direction of international traffic.

Of course, as far as pilots were concerned, the combined effect of an overweening state and stricter international regulation was far from a bad thing. Certainly those who were fortunate enough to work for the growing number of state-subsidised and nationally and internationally-regulated flag carriers (such as the British Overseas Airways Corporation (BOAC), British European Airways (BEA), Air France and Pan American World Airways) enjoyed a reasonably comfortable working environment and lifestyle. The flag carriers embodied the values of the state that spawned them. Flag carriers reified (made tangible) the motherland's defining qualities. They were a

11 The Republicans also intervened. President Hoover, for example, '... instructed the biggest US airlines to *plan* a network. That is how American Airlines ..., Trans World ... and United Airlines ... acquired their distinct east-west territories, while Eastern Airlines ... served the Atlantic zone' (Reed 2002, p. 5).

distillation of their paymasters' history, culture, values and aspirations. Their Captains and First Officers were, in all but name, pilot-civil servants, dispatched across the globe to serve and (subtly) promote the national interest. According to Hudson and Pettifer (1979, p. 129) flag carriers were seen by sponsoring governments '... as a great opportunity to wave the flag at foreigners. The expense was a secondary consideration'.[12] Bennett (2002b, p. 7) notes: 'A flag carrier is national ambition made tangible. It is a *political instrument*'. If a flag carrier is a political instrument, then its workers are covert political agents (note the small 'p' for political, however).

During their early days, Britain's flag carriers, BOAC and BEA, seemed to exist solely for the purpose of transporting the country's political elite – at considerable cost to the taxpayer. Ted Williams, an ex-BOAC navigator, notes: '[O]nly a tiny number of people ever flew with us – and they were almost all diplomats, or Senior Service officers. We estimated that it cost two and a half times what BOAC was actually earning to carry each passenger' (Williams cited in Quinn 2003, p. 63). Taffy Barrow, an ex-BOAC pilot, comments: 'I flew Hythes [the Hythe was a flying boat] to Australia and the Far East .... most [passengers] were civil servants and army people' (Barrow cited in Quinn 2003, p. 178). To understand what it was like to work for a flag carrier before deregulation the next section describes pilots' experiences of three national airlines: BOAC, BEA and Pan American. (Apologies are offered in advance to the Captains, First Officers, engineers and navigators of the many great flag carriers that, perforce, are not surveyed.)

### *Pilots' experiences of BOAC, BEA and Pan American*

BOAC and BEA were set up in 1939 and 1946 respectively. BOAC provided long-range international services while BEA provided European and domestic services. Gaskell (1999, p. 9) notes: 'Both of these State-owned airlines ... evolved during the post-War years under conditions of considerable political control and interference ... '. Political interference created tensions between government and the flag carriers' managements, especially in the matter of aircraft purchases (Gaskell 1999, pp. 13–14).

Flying for one of Britain's two flag carriers had both positive and negative aspects. On the positive side, flying in the 1940s and 1950s was still a relatively exclusive activity that put the flight crew, and especially the Captain, on a social pedestal. Taffy Barrow notes: 'The old Imperial Airways captains I worked for thought of themselves as very grand – as lords of the air, you might say – and they were very conscious of their importance and status. They were also very concerned that the crew should treat them with due respect ... ' (Barrow cited in Quinn 2003, p. 173). Beaty's (1969, p. 16) analysis is similar to Barrow's: 'Early civil airline pilots certainly considered themselves something special, and insisted on a standard of subservience from their company ground staff which may have left permanent

12 Flag carriers were more benign instruments of diplomacy than gunboats or army expeditionary forces.

marks on the memories of those who eventually rose a good deal higher than pilots in airline hierarchies'. The fact that airline managers suffered bruising encounters with imperious Captains during their early careers may have coloured their outlook and subsequent behaviour. They might have been unnecessarily subservient to flight crew, or unnecessarily harsh. Extremes of behaviour are not conducive to good industrial relations.

Cabin crew were generally respected by their charges, as Hazel Rose, an ex-BOAC stewardess explains: 'The passengers always seemed to be really nice to us ... none of that rudeness you hear about today when flying is such a pressurised business' (Rose cited in Quinn 2003, p. 140). Olive Carlisle, another ex-BOAC stewardess, paints a similar picture: 'Passengers were always very friendly, but I suspect this had a lot to do with the fact that they were rather in awe of us because flying was not by any means something you took for granted and we were the people who made it possible' (Carlisle cited in Quinn 2003, p. 83).

It could be argued, of course, that the deference noted by these crewmembers was just another manifestation of the general social deference of the day. Britain in the 1940s and 1950s was a country in which most citizens deferred to authority figures like teachers, doctors, scientists, policemen and religious leaders and to institutions like the Monarchy, the Church of England, Parliament and the British Broadcasting Corporation (BBC). This deference migrated across international borders. As Ted Williams, an ex-BOAC navigator remembers: 'Another difficulty we experienced was with air traffic controllers at Karachi: they would never contradict us if we were wrong when we gave our position – they always simply assumed we must be right, which is a very dangerous thing for an air traffic controller to do. If you asked for a bearing they would simply try to give you the bearing they thought you thought you were on .... Perhaps it was simply the habit of deference to the British! But anyway, you needed nerves of steel to fly into Karachi when they were in the mood to agree with everything!' (Williams cited in Quinn 2003, p. 60).

Some of those who worked for the flag carriers came from Britain's social elite. This resulted in a rather exclusive employee social profile that may have induced an even greater display of deference on the part of passengers and foreign agents (like Karachi's air traffic controllers). Hazel Rose (whose father untypically worked as a furnaceman in the steel industry) notes: '[I]t was common knowledge that stewardesses were always wealthy and very well educated .... They probably didn't need to work for the money ... but flying was glamorous and rather chic ... which is what appealed to them' (Rose cited in Quinn 2003, p. 134).

As for the pilot community, it too had a rather exclusive social profile. One wartime pilot who had been obliged to leave school at 14 but who went on to gain his commission and fly for BEA in the 1950s comments: 'In England with my background [Vic Avila worked as an engine fitter for a tractor firm] I would undoubtedly have been made a sergeant pilot ... [but] in Canada the old school tie counted for nothing and on the basis of my ability I got [a commission]. So two years from being accepted as a pilot I was a pilot officer. Some of my colleagues complained bitterly about this – they didn't like the fact that I had been granted a commission when I hadn't had their education' (Avila cited in Quinn 2003, p. 45).

It could be said that British commercial aviation in the 1940s and 1950s involved one group of relatively privileged professionals (flight and cabin crew) ministering to another group of relatively privileged professionals (passengers travelling abroad for business or pleasure). Vic Avila notes: 'Although passenger flying had grown considerably since the 1930s and 1940s it was still something pretty much the preserve of the rich or for people whose tickets were paid for by their employers' (Avila cited in Quinn 2003, pp. 47–48). Hudson and Pettifer (1979, p. 131) comment: 'The truth was that the international air transport industry [of the 1950s] was being run on a quite unnecessarily lavish scale. With State funds always available to meet the annual deficit there was no incentive to be economical, no reason to adopt measures which would make the business pay'.

The lack of incentive to economise (despite the austere economic conditions of the 1940s and 1950s) meant that flight and cabin crew enjoyed a relatively comfortable lifestyle – at least when off duty. Ted Williams, an ex-BOAC navigator, recalls: 'I suppose it was quite a glamorous life because we did get the chance to stay in Cairo or Basra or wherever while we waited for our next aircraft. You'd slip – or change – crews, but not on every leg of a journey. You might slip for a week – and this could be very pleasant. In Cairo, for example, BOAC had a luxurious Nile cruiser for crew accommodation' (Williams cited in Quinn 2003, p. 63). Olive Carlisle, an ex-BOAC stewardess, states: '[We enjoyed] comfortable nights in luxury hotels across the world' (Carlisle cited in Quinn 2003, p. 83). Hazel Rose remembers: '[I]n those days we'd fly to Rome and stop for two days, then to Beirut where we'd get another two days, then Karachi and two days there. Today you fly to Rome and turn round and fly back – we used our two days to explore and we were paid an allowance to do it. It's only now I realise how privileged we were and how much fun we had! The stewardess's job has changed completely in this respect – much more pressurised and less time to enjoy anything. We didn't always stop for two days but it was never less than a day. Sometimes three to five days. In Beirut they put us up in the Bristol Hotel which was wonderfully luxurious ... ' (Rose cited in Quinn 2003, pp. 139–140). Len Reddington, a wireless operator, recalls: 'Being a wireless operator on civilian aircraft was very different from my old RAF days. For a start we got to stay in very good hotels all over the place' (Reddington cited in Quinn 2003, p. 161).

Given that a flight to Australia in the 1950s in a BOAC Lockheed Constellation 749 (a relatively modern piston-engined aircraft) took many days, flight and cabin crew were able to enjoy good hotel accommodation, interesting food and some quite flamboyant lifestyles for weeks at a stretch. The contrast with the lifestyle back in Britain could not have been more stark. Britain in the 1940s and 1950s was a 'monochrome' society. Almost bankrupted by the Second World War the country had very little money to spend on itself. The nation's diet was as monotonous as its weather. Its cities were drab, its people subdued. Hennessy (1993, p. 276) observes: 'An especially sharp blow to public morale came in the summer of 1946 when bread was rationed. It hadn't happened in the war'. In 1948, Britain hosted the Olympic

Games. Childs (1995, p. 95) notes: '[C]ompetitors from the more affluent countries were surprised at the austerity they experienced in London and the ruins of bombed buildings'. According to Childs (1995, p. 95), the 1951 Festival of Britain 'was ... regarded as a morale booster'. One can only wonder what the long-suffering British public thought of BOAC and BEA flight and cabin crew, whose comparatively glamorous lifestyle they were subsidising through the Treasury.

Of course there were some negative aspects to flying for BOAC and BEA in the 1940s and 1950s, amongst them some very long duties. Ted Williams recalls: 'We ... flew hours that would terrify today's regulators. You might easily do sixteen hours at a stretch ... ' (Williams cited in Quinn 2003, p. 59). Stewardess Olive Carlisle remembers working fifteen and a half hours on a return trip from Nairobi to Cairo with just two thirty-minute stopovers (cited in Quinn 2003, p. 75). BOAC pilot Taffy Barrow was flying C-class Empire flying boats between Durban and Calcutta. Quinn (2003, p. 174) notes: 'Durban to Calcutta and back took five weeks, and in those five weeks pilot and copilot would be at the controls for as much as 220 hours'. Compared with the automated aircraft of today, the C-class Empire flying boats required a considerable input from their pilots. Even the Comet, BOAC's first pure-jet aircraft, required close attention, as Vic Avila explains: 'Even with the Comet there was a lot to do ... compared to a modern plane where you have a smaller crew and more automatics' (Avila cited in Quinn 2003, p. 51).

Of course, senior crewmembers were in a position to delegate to junior crewmembers. Some took advantage of the command gradient, as Taffy Barrow notes of his C-class captains: 'As soon as we took off the captain would disappear to somewhere more comfortable and you – as first officer – had to fly the plane, monitor all the instruments and so on. You were also first on at five in the morning and last off at night' (Barrow cited in Quinn 2003, p. 173). There were compensations, however, as Barrow explains: '[W]e always stopped somewhere to sleep at night. It was truly leisurely and luxurious by the standards of later passenger flying' (Barrow cited in Quinn 2003, p. 174). The relatively short range of the C-class meant that it was airborne for no more than about four hours at a time. Today, of course, ultra long-haul aircraft like the Airbus A340-500 might be in the air for anything up to eighteen and a half hours (although shift changes mean that flight and cabin crew would not be on duty for the entire flight) (Clark 2004a).[13]

13 In February 2004 Singapore Airlines (SIA) launched a non-stop service between Singapore and Los Angeles. The outbound service takes 16 hours. The return service takes even longer. According to Clark, SIA was planning to launch an 18 hour direct service between Singapore and New York. In publicising the service, SIA's Senior Executive Vice-President focused on the advantages to *passengers*, and especially to business travellers of ultra long-haul: 'With the A340-500, we are offering non-stop services over long distances, that translate to greater convenience, comfort and time savings. It is especially designed for today's business travellers' (Tan cited in *Skytrax Research* 2005). The *Skytrax* report did not discuss the convenience and comfort of SIA's flight and cabin crews. The report ended with this summary from SIA's Senior Executive Vice-President: 'The interior of the A340-500 – from the cabin layout to the seats – has been designed to offer excellent comfort for long-

In his 1984 book *Empires of the Sky*, Sampson argued that Britain's 'chosen instrument' for long-haul travel had seen four ages: 'Over ... six decades Britain's "chosen instrument" [has] undergone every kind of metamorphosis: it [has] been a pioneers' airline, an imperialists' airline, a pilots' and engineers' airline, and now it [is] a marketer's airline' (Sampson 1984, p. 219).[14] Whether BOAC in the 1960s and early 1970s could be described as a pilots' airline is open to debate. While pilots worked reasonable hours they were paid less than their colleagues in other airlines. Dienel and Lyth (1998, p. 71) explain: 'A strike by BOAC pilots in the summer of 1968 drew attention to the fact that BOAC used them less, and paid them less, than other airlines. Pan American pilots received three times the salary of their BOAC colleagues for flying identical aircraft, but they flew 960 hours per year compared to the BOAC average of 550 hours. The British pilots were actually willing to work longer hours, but management was not ready for them to do so – it was perhaps one of the few strikes on record in support of a claim for longer working hours'. Both BOAC and BEA suffered from low productivity. This was partly a product of low aircraft utilisation (a problem confronted head-on by today's LCCs): 'Like BOAC, BEA suffered from high operating costs .... [L]ike BOAC ... it ... had generally low aircraft utilisation and labour productivity. In 1959 BEA's total operating costs were about 45 per cent higher ... than American domestic airlines ... despite the higher wage levels in the United States' (Dienel and Lyth 1998, p. 69). Perhaps it would be more accurate to describe BOAC in the 1960s and early 1970s as a pilots' airline in terms of flight and duty times, but not in terms of remuneration.

The 1960s proved to be a political watershed for Britain. This was the decade when Britain's political elite (if not the British people) came to accept that the Days of Empire were finally over and that state enterprises like BOAC and BEA, freed from their role as the logisticians of the Foreign Office, could become more commercial. The façade finally cracked in 1959. Dienel and Lyth (1998, p. 68) describe 1959's rupture with the past thus: 'In 1959 BOAC faced severe criticism in the House of Commons when successive speakers claimed that the airline was poorly managed and that its earnings on capital were low. BOAC's aircraft were fast but were not worked hard enough .... There was a general view that the airline needed more tourist class passengers and had to speed up "its rate of conversion from a high quality service operator to a mover of mass traffic"'. Parliament's questioning of BOAC's role was prescient. On February 3 1960 the Prime Minister, Harold Macmillan, made his now-famous 'Winds of Change' speech in Cape Town, South Africa. He stated: 'The wind of change is blowing through this continent and our national policies must take account of it'. Macmillan was publicly accepting what the South African

haul business travellers to ensure they arrive refreshed and ready to go'. One wonders how 'refreshed and ready to go' flight and cabin crew will be on arrival?

14 Note that the Thatcher Government privatised British Airways in February 1987.

Parliament could not accept – that the Days of Empire were drawing to a close.[15] The implications for BOAC were obvious.

In its appeal for greater economic efficiency the House of Commons picked up on a major new social phenomenon – consumerism. Hudson and Pettifer (1979, p. 157) observe: '[C]onsumerism ... began to be very noticeable in the air travel world during the 1960s. Today we take criticism of airlines for granted, but, on a large scale, it was something new in the 1960s'. As mentioned above, Britain after the Second World War was a determinedly deferential society where the excesses of the Establishment were generally brushed under the carpet. That all ended when a British Minister of the Crown, John Profumo, compromised national security by having sexual relations with a woman who was also involved with an official from the Soviet Embassy. The British press had a field day. Profumo resigned. Macmillan's government was defeated at the next General Election by a party led by a Grammar School[16] boy, Harold Wilson. The Days of Deference were numbered. British society was never quite the same again. (Today, of course, everyone is fair game for the British media, from Prime Ministers (and their wives) to Princesses to pilots.)

British politicians' desire to make BOAC less elitist was very much in tune with the 1960s' zeitgeist of debunking venerable institutions like the Church and the Monarchy. If 1950s' Britain was a country transfixed (paralysed?) by class, deference and historic obligation, 1960s' Britain was a country transfixed by the future. Harold Wilson's two Labour governments of the 1960s invested significant financial and political capital in Britain's capacity to innovate.[17] Gilbert (1999, p. 314) explains: 'In Britain a new Labour government ... pledged to move Britain forward into the age of modern science and technology'.

Britons were wealthier in the 1960s. They were better educated (and were therefore more inclined to ask 'inconvenient' questions of authority figures like politicians, doctors, teachers and clergymen, and less inclined to conform to the norms of yesteryear). They had more leisure time. They wanted to explore. They demanded affordable air transportation. Against this background of slow-burn social revolution, institutions such as BOAC and BEA seemed anachronistic and, as far as an increasing number of Britons were concerned, irrelevant. Many looked to the charter and inclusive tour (IT) operators to meet their air transportation needs.[18] The United Kingdom's chosen instruments had to change.

---

15 Macmillan's analysis was correct. During the 1960s numerous British colonies, including Kenya, Ghana, Sierra Leone and British Somaliland, negotiated their independence from the United Kingdom (Gilbert 1999, pp. 223–225).

16 Grammar Schools were publicly-funded selective schools. They were a poor man's public school.

17 The resulting flagship projects included fast-breeder reactors (now abandoned), the Severn bridge (bold and elegant, but, in light of traffic growth, an under-engineered structure) and Concorde (simultaneously a technological marvel and economic disaster).

18 Maslen (2003, pp. 123–124) notes: 'With an upturn in the global economy, disposable household income increased significantly during the 1970s and demand for foreign holidays

Across the water, Pan Am, like BOAC/BEA, experienced mixed fortunes during the Twentieth Century. Like BOAC/BEA, Pan Am was America's chosen instrument – its flag carrier. As with BOAC/BEA, Pan Am's fortunes fluctuated with those of its country of origin: The Second World War transformed the economy, outlook and international stature of the United States. As Ambrose (1985, pp. xiii–xv) explains: 'In 1939 ... the United States had an Army of 185,000 men .... America had no military alliances .... The dominant political mood was isolationism .... At the conclusion of World War II America was at the top of the mountain. In all the world only the United States had a healthy economy, an intact physical plant capable of mass production of goods, and excess capital .... During World War II, Henry Luce of *Life* magazine spoke for most political leaders ... when he said that the twentieth century would be the American century'. As America's chosen instrument, Pan Am, and one Juan Trippe, had a golden opportunity.

During the 1930s, Juan Trippe, a Yale-educated, talented, stylish and well-connected patrician, created an airline in his own image. As Sampson (1984, p. 46) observes: 'Trippe cultivated the style of a public service, putting his pilots and crews in naval-type uniforms .... Pan Am's overseas monopoly, and its lack of domestic land routes gave it a more dignified image than domestic airlines .... Trippe's airline was already being widely attacked by South Americans as an adjunct of the State Department, an instrument of American imperialism'.

Pan Am, like its president, saw itself as part of the American establishment – steady, dependable, patriotic. Stylish, even. Gandt (1999, pp. 3–4) offers this analysis of Pan Am's self-image and outlook in the 1960s: 'Each [new hire pilot] knew he could have gone to another airline. There was Delta, for example, if you wanted to live in the South. There was Braniff, Texas-based and expanding very fast .... But those were *domestics*, like Greyhound Bus. This was *Pan American*. No bus driving for them. No miserable lunches in places like Des Moines or Cincinnati or Boise. For them it would be sushi in Tokyo, *petite dejeuner* in Paris, tea in London. Pan Am advertised itself as "the world's most experienced airline". Well, sure, thought the new hires. That was the kind of airline they *ought* to be working for. Pan American was the world's most glamorous airline. It was also the most snobbish'.

According to Gandt (1999, p. 209) Pan Am's pilots withdrew their labour only with the greatest reluctance, such was their loyalty to the *ideal* of Pan American World Airways: 'A strike at Pan Am was as rare as a visitation by aliens .... [A]t blue-blooded Pan American, the Imperial Airline, a strike was viewed with the same distaste as getting sucked into a bar fight. It besmirched the airline's image. There was something about it that just wasn't ... *imperial*'.

Pan Am's track record in the 1950s and 1960s spoke for itself. The more self-aware pilots realised how fortunate they were. The less self-aware pilots only realised this when they were furloughed (there were two major furloughs: 1969 and 1976). As Gandt (1999, p. 223) explains: 'Until the day of the F word, they had

---

rose. As one of the pioneers of the charter market in Europe, UK-based Britannia had offered inclusive tour packages to the Mediterranean in the mid-1960s ... '.

lived their entire adult lives beneath an institutional umbrella – home, college, the military, then the airlines. Not one minute of their lives had been spent languishing in unemployment lines or perusing classified job ads'. In other words, until the economic perturbations of the 1970s and early 1980s (when 'Reaganomics' began to bite) Pan Am pilots had led a *privileged* existence – a fact that had not gone unnoticed by at least one United States Air Force transport pilot: 'Chuck Kraft ... had spent the past four years eating box lunches and drinking bug juice and flying Air Force C-130 transports [i.e. Lockheed's C-130 Hercules turboprop transport aircraft] around the world. During his missions a scene had begun to repeat itself. At each layover – Madrid, Paris, Bangkok, Hong Kong – he would head directly for the hotel where the airline crews stayed. At the swimming pool he would observe the same curious sight: *Pan Am Stewardesses* – Swedish, German, French, American – gorgeous in their bikinis, frolicking around the pool. Frolicking with them would be smiling, suntanned, obviously overpaid *Pan Am pilots.* One day in Bangkok, standing there in his evil-smelling, baggy flight suit ... Kraft suddenly fathomed an immense truth: *They're doing the same goddamn job I am*' (Gandt 1999, p. 4).

It might have escaped Kraft's attention at the time, but those smiling, suntanned and 'obviously overpaid' Pan Am pilots were in a vulnerable position. When the industry was buoyant – as it was for much of the 1960s – pilots were in demand. Life was good. The cars were fast. The hotel pools were full. But when the industry caught a cold, as it did on more than one occasion in the 1970s, those same, hitherto much-in-demand pilots could find themselves looking for alternative employment.[19]

### *Commercial aviation's 'grunts'*[20]

Each mode of transport has its glamorous and not-so-glamorous sectors. During the post-War period, Britain's merchant marine, for example, operated luxurious ocean-going liners alongside humble workhorses like the Clyde Puffers (small coasters). British Railways operated the magnificent Merchant Navy class alongside its stalwart but resolutely unglamorous Black Fives. Aviation also had its less glamorous corners. Not all pilots flew for heavily subsidised, globetrotting flag carriers like BOAC. Some flew crop dusters. Others air taxis. Others lumbering freighters. Often the aircraft flown by the grunts were pensioned-off passenger aircraft – airworthy but hardly state-of-the-art (see, for example, Gandt's, 1999, p. 224, comments, below).

Although unglamorous, freight operations have promoted economic growth, facilitated the exploitation of natural resources (see Hudson and Pettifer 1979, pp.

19 Dean (cited in Pirihi 2005) comments: '[N]ot many pilots retire early because of the high earnings and benefits, and the fact their unique skills are not generally transferable to other careers'.

20 During the Vietnam War, US troops were referred to as grunts because, despite the hardship and danger, they got the job done. They could take it ... the 'it' being razor-sharp elephant grass, poisonous insects, extended patrols with no sleep, rotten food and frequent ambush by an unseen yet knowing enemy. US director Oliver Stone got close to the truth of the grunt in his film *Platoon.*

171–192) and, in the guise of the 1948 Berlin Airlift, thwarted Soviet geopolitical ambitions. In June 1948, the Soviets blockaded Berlin in an effort to dislodge the Allies from the western half of the city. The West decided to supply the city's two and a half million citizens by air – a remarkable decision in light of the very basic technology at their disposal (chiefly Douglas Dakotas). A city the size of Berlin required 5,000 tons of fuel and food each day. The Labour Government's Ernest Bevin contested: 'You could never feed two and a half million people by air'. But the air forces of Britain, France and the United States, aided by an armada of chartered civilian aircraft, did exactly that.

Many of Britain's fledgling charter companies provided aircraft and crews. Those involved included Airwork, Eagle Aviation and Flight Refuelling. It was a baptism of fire. Crews flew until they were exhausted. Aircraft were flown until the rivets popped (whereupon they were re-riveted and returned to the line). But it was a cathartic experience for Britain's private air services sector. As Eglin and Ritchie (1980, pp. 16–17) explain: 'The Berlin airlift was the making of British independent airlines .... The generous charter rate and the huge number of hours flown produced enough cash flow for the independents to set up complete organisations, with flying crews, ground staff, UK bases, plant and equipment. And the unrelenting tempo of the airlift, with month after month of round-the-clock flights, taught the independents a lot about how to run a continuous air operation'. According to Eglin and Ritchie (1980, p. 18) the Airlift transformed a disparate group of freewheeling contract pilots into a coherent, purposeful team: '[The Airlift] generated a spirit of teamwork that welded the entrepreneurial, freebooting sky tramps into teams as closely knit as the RAF and USAAF regulars'.[21]

The fact that the private air services sector came of age during the Berlin Airlift did not change the British Government's relationship with its flag carriers, however. BOAC and BEA continued to receive heavy subsidies. Their pilots were still regarded as an elite. They were, after all, Britain's 'flying ambassadors'. Meanwhile, companies like Freddie Laker's Aviation Traders continued to operate out of unglamorous tertiary airfields such as Southend (this is the way many of today's LCCs operate, of course). Laker's grunts worked hard: '"Work for Fred and burn your bed" was one slogan scrawled on the operations notice board at Hamburg' (Eglin and Ritchie 1980, p. 17).

Despite private contractors' appetite for hard work, when big contracts like the Berlin Airlift ended, life got tough. Unlike their glamorous cousins (the flag carriers) they had no Minister of the Crown to bail them out. Consequently: '[W]hen the airlift ended, all the embryonic airlines which had sprung up found themselves unemployed. Many of them collapsed almost overnight' (Eglin and Ritchie 1980, p. 21). Redundant pilots had a choice: look for a job in the UK, or move on. In aviation it paid to have few ties.

21 Of course many grunts had flown for the military during the War. Consequently they were familiar with military discipline and teamworking.

*Flying into the Swinging Sixties*

Commercial aviation expanded rapidly during the 1950s and 1960s. As Taylor (1967, p. 124) explains: 'Air travel grew at a fantastic rate. Before long more people were crossing the Atlantic by air than by ship, and the total number of air travellers became astronomical. In 1945, about 9,000,000 flew on scheduled airline services all over the world, excluding Russia and China which published no figures. The total twenty years later had risen to nearly 200,000,000, of whom more than 4,000,000 flew over the Atlantic'. Needless to say, trans-Atlantic carriers such as Pan Am and BOAC, and the pilots who were fortunate enough to fly for them, benefited from this growth. Then came the decade of retreat – the 1970s.

In the USA, the prolonged agony of the Vietnam War combined with the acute shock of the 1973 October War served to undermine confidence in the economy. The ready availability of unlimited quantities of cheap oil during the 1960s had underwritten America's geopolitical ambitions. The USA had been in a position to spend enormous sums of money on a military adventure in south-east Asia. Hundreds of thousands of troops were dispatched. New weapons were developed. Additionally it had spent huge sums on the Apollo space programme (which in scientific and political terms had been broadly successful). Nixon, perhaps seeking to emulate his predecessor Johnson's affection for grand projects, had even launched a War on Cancer.[22]

While the success of the space programme served to boost morale, the gradual realisation that the US Military had failed to halt the spread of communism in south-east Asia undermined national confidence. As Ryan (1995, pp. 103–105) explains: 'US commitments during the Vietnam War resulted in massive costs, which could not easily be justified. Most pertinently, the loss of 58,123 American lives, but also the opportunity costs of President Johnson's Great Society programme for social restructuring and the economic costs in the international context .... After 1968 the so called "Vietnam syndrome" exerted an influence on the will to use power at a time when it was more necessary to do so due to the relative economic decline of the United States'.

There is broad agreement that the 1973 Middle East conflict and subsequent quadrupling of the price of crude oil caused significant problems for the west. Gilbert (1999, p. 467) comments: 'As a result of the sudden and precipitate rise in the price of oil, there was massive inflation, an economic depression, and, as the cost of fuel oil soared, a drastic reduction in petrol supplies'. In Britain the failure to invest and mounting industrial unrest added to the country's economic woes. As Morgan (1993, p. 581) explains: 'In the nineteen-seventies, Britain offered a permanent, painful case for macro-economic and sociological treatment. Its economic decline

---

22 President Johnson's ambition was to build The Great Society – a United States of America underpinned by ideals of social justice and equality of opportunity. The grinding poverty he had seen in home regions like West Virginia, the paradox of want in the midst of plenty, had scarred him.

continued, comparatively in relation to almost all other developed nations, and even in absolute terms compared with earlier production levels. It was much aggravated by the dramatic change in the energy situation in 1973–4 as a result of which Britain and other Western nations suffered a fourfold increase in the price of Middle East oil .... [T]he huge rise in the price of oil inevitably fuelled inflation on a scale unknown since 1919. It was reinforced by trade union pressure for enormous wage increases of anything up to 30 per cent .... British inflation continued to run at a historically high level .... With surging price rises and pressure from wage and other unit costs, unemployment re-emerged as the major domestic scourge. By 1980 it was over two million, a total unknown since the thirties .... [U]nemployment had advanced to well over three million by the spring of 1983. There seemed to be a deep rot at the heart of the economy ... '. Theobald (1994, p. 67) notes: '[T]he 1979 increase ushered in a second recession which ... proved to be longer-lasting .... In Britain ... unemployment in the eighties topped 11 per cent, more than twice the highest figure for the seventies (4.8 per cent in 1977)'.

Britain's high unemployment levels, falling birth rate, ageing population and commitment to monetarism (that is, fiscal prudence) had two major implications. First, the productivity of those left in work had to increase. Secondly, given its preference for low levels of taxation the government had no choice but to plunder its coffers to meet its welfare obligations. In order to shift resources (that is, move money to those areas of the economy where it was most needed) the government's role in the wider economy had to be curtailed. Theobald (1994, p. 68) observes: '[T]he determination to "roll back the state" ... heralded a new era in the organisation of work'. The stage was set for a remodelling of the social market economy and for the privatisation of assets like British Airways (the product of the 1974 merger of BOAC and BEA).

On the other side of the Atlantic the economic perturbations of the 1970s were having a major impact on commercial aviation. Of most concern to pilots was the fact that major carriers like Pan Am were releasing flight crew. Some Pan Am pilots found positions with other, less glamorous carriers. As Gandt (1999, p. 224) explains: 'Some of the furloughees found jobs with the nonscheds – fly-by-night charter operators with cast-off Boeing 707s and Douglas DC-8s – hauling freight and cattle and produce up from Latin America and out to the Middle East. The pay was lousy. The airplanes had peeling paint and anaemic engines. But the furloughees grabbed up the nonsched jobs, when they could get them. It was flying'. Those less lucky had to settle for non-flying work: 'Some ... tried selling insurance and used cars. Some ... built houses. A few [went] into construction work ... ' (Gandt 1999, p. 225).

The fate of Pan Am's furloughees in the 1970s raises an interesting question, namely: 'What opportunities are open to career commercial pilots when they are made redundant (or when they are retired or released on medical grounds)?' Many of Pan Am's pilots had just one skill-set – the ability to fly a commercial passenger aircraft. Through no fault of their own Pan Am's furloughees constituted an inflexible labour resource. Some had the inclination and money to retrain. Those who were unwilling or unable to acquire a new skill-set had to take whatever job they could get, however menial.

Interestingly, many furloughed pilots – however successful they had been in their new career – returned to flying as soon as the opportunity presented itself. As Gandt (1999, p. 225) recalls: '... Pan Am start[ed] calling back the pilots it had furloughed ... they were glad to be on the payroll .... It was enough just being in the cockpit, sitting sideways in the flight engineer's seat of a beat-up 727 .... Some had established successful new careers. They were coming back to Pan Am because during all those years they thought they were missing something. Flying was something they loved'. It is probably true to say that flying has something of an addictive quality, and that this quality informs whatever decisions a pilot makes throughout her/his life. *Flying gets under the skin.* Industry bosses may also be susceptible to the 'flying bug'. As the Chairman of Air Florida remarked on one occasion: 'Once you get hooked on the airline business it's worse than dope' (Acker cited in Sampson 1984, p. 137).[23]

*Finally, deregulation*

Conscious of the need to reinvigorate US commercial aviation, the Carter administration turned to the market. Confident that free markets could deliver a better level of service to the consumer than regulated markets Jimmy Carter set about dismantling US domestic aviation's regulatory straightjacket. The seeds were being sown for commercial aviation's 'roaring, deregulated Eighties' (to borrow Gandt's, 1999, p. 266, phrase) and for deregulation's gradual migration across the globe. (In 2006, deregulation was at work in countries as diverse as India and China. The deregulation tide, like economic and cultural globalism, would appear to be unstoppable.)

Deregulation had two major effects. First it benefited consumers. Services were cheaper and there were more of them. Communities that had not been served by the major carriers found themselves being served by start-ups. Secondly, it had a negative impact on the majors – US commercial aviation's 'establishment'. As Sampson (1984, p. 138) explains: 'Most of the big airlines were in serious difficulties after three years of deregulation and recession. While the local airlines were able to maintain profits ... the airlines on trunk routes were showing mounting losses ... '. Before deregulation, says Sampson, airlines like Pan Am saw themselves as venerable institutions. They believed themselves to be deserving of respect. They were sure the public would not desert them. After all, had they not served the nation dutifully in

23 In a spirit of 'verstehen' (see Chapter 1) the author wishes the reader to note that he, too, is addicted to flying. The addiction began some six years ago during a cloudless, late-night sector from Edinburgh to Nottingham East Midlands, during which the whole of southern England was visible, lit up like a Christmas tree. Cities appeared as pools of light. The author got up from the jump seat for a better view. Standing there on three millimetres of aluminium in an aircraft travelling at 75% the speed of sound five miles above the earth, the flying bug bit – an amalgam of pride in the technology, awe and sheer excitement.

peace *and* war?[24] After deregulation, argues Sampson, it became obvious that the public, as consumers in a free market, were interested only in cost and convenience: 'The big corporations like Pan Am ... had seen themselves in the Fifties and Sixties as historic institutions which could command a special loyalty from their passengers. But deregulation soon revealed that most passengers only cared about two things – price and convenience' (Sampson 1984, p. 140).

This focus on price and convenience would seem to be a perennial trait of consumers of aviation products. In 2005 a new regional airport (built on the site of the abandoned RAF station at Finningley near Doncaster) opened in direct competition with Nottingham East Midlands Airport (NEMA). NEMA's management was unperturbed. They told the press they believed their customers would remain loyal. A television news team interviewed some of NEMA's customers inside the airport's main terminal building. They were asked whether they would remain loyal to NEMA regardless of the new local competition? Most responded that their choice of airport would be determined by two factors only – the services operated from the airport, and the cost of those services. It was apparent that sentiment would play no part in the decision-making of most interviewees.

### *Consumers' experience of deregulation*

The deregulation of the 1980s and 1990s changed the social, economic and political complexion of commercial aviation. Airline deregulation was a symptom of a much wider push to re-introduce laissez-faire capitalism – that form of social, economic and political organisation that had fuelled the Industrial Revolution of the Eighteenth and Nineteenth Centuries. As Shearman (1992, p. 83) explains: 'The evolving global political and social scene is largely and increasingly dominated by Western capitalist thinking. In particular this involves the right of businesses to trade freely in goods and services in domestic markets with rewards being decided by market demand, i.e. the principle of "laissez-faire"'.

Reaganomics, Thatcherism and the various other manifestations of the 'new laissez-faire' saw the privatisation of some flag carriers (like British Airways) and the cutting of subsidies to others. Barriers to market entry were lowered or eliminated. Start-up airlines proliferated. Operators like People Express in the USA sought to meet the travel needs of the Greyhound Bus generation – those who wanted to 'hit the road' Kerouac-style, but who preferred to spend their hard-earned money on comforts (like a room with a bath and one good meal a day) rather than fares. Being independent-minded these customers had never taken to the inclusive tour

---

24 With regard to Pan Am's self-image Petzinger's (1995, p. 187) comment is apposite: 'Pan Am [was] a company practically frozen by the inertia of grandeur and tradition'.

(IT) market. IT operators did all your thinking for you.[25] That wasn't the style of the Greyhound Bus generation. That's not the way Kerouac would have done it.[26]

Start-ups like People Express saw the airline industry in an *entirely* new way. As Sampson (1984, p. 136) notes: 'Donald Burr ... founded his own airline People Express with an almost messianic zeal for his own version of capitalism .... He had no romantic illusions about airlines. "They're a low-end commodity", he said. "They're capital-intensive, labour-intensive, fuel-intensive. You can't sell the seats with sex or caviar: the highest value is the price"'.[27] (Of course prior to deregulation, when the airlines had no choice but to charge the fares set for them, the only way they *could* differentiate their products was through glamour, food, paint schemes and other frivolities.)

According to Shearman (1992, pp. 84–89) the impact of deregulation in the US where '... the ... air transport market is particularly large with highly sophisticated travellers' has been 'enormous'. The number of services increased, while their cost reduced: '[T]he town which lost its once-a-day Boeing 737 service gained, perhaps, a twice-a-day Shorts 360 service instead .... The industry [saw] the advent of new airlines with low overhead costs or with non-unionised labour .... The result not surprisingly was that air fares fell. Compared with 1974 the average air fare in real terms in the mid 1980s had fallen by 25% ... passenger traffic grew rapidly ... the number of domestic passengers ... rose from 173 million in 1971 to 457 million in 1987, an average annual increase of over 10%'.

The traditional airlines (that had resisted deregulation) 'had to lower their costs and increase their productivity to cope with the lower yields arising from the cheaper fares they were forced to offer to match competitors'. Shearman concludes: '[T]he industry has changed – in fact it is almost unrecognisable from the cosy system that existed some 15 years ago'. Morrison and Winston (1995, p. 159) concur: 'As deregulation has evolved, the U.S. airline industry has become more competitive, generally providing travellers with lower fares and more frequent service. Although industry leadership has changed hands from the established carriers to new entrants, back to the established carriers, and lately to low-cost carriers, it has become increasingly clear that airlines cannot count on exploiting opportunities to charge and maintain high fares to remain profitable .... As the industry evolves ... it will take appropriate measures to lower costs and stabilise revenues without raising real

---

25 As Williams (2002, p. 171) explains: '[IT products are delivered] by [c]arriers that are owned by vertically integrated tour operating companies [that] incorporate travel agencies, hotels and ground transportation companies'.

26 British entrepreneurs like Freddie Laker (later joined by Richard Branson and Stelios Haji-Ioannou) also realised there was money to be made from the new generation of independent-minded, confident, well-educated, uncommitted freewheelers.

27 Capital and labour-intensive industries like commercial aviation and steel-making (the classic 'sunset' or 'rust-belt' industry) stood in sharp contrast to sunrise industries like software production, which were fuelled largely by intellect and insight, and which could be sustained by a relatively small number of people (often working from home through the medium of broadband services).

fares or compromising safety. In short, although the ride has hardly been smooth, the industry is moving in the right direction'.

*Pilots' experience of deregulation*

While US consumers benefited from deregulation, US pilots did not. The coincidence during the early 1980s of deregulation and recession served to undermine pilots' hard-won gains. As Sampson (1984, p. 142) explains: 'Over fifty years the pilots had built up a position which other professional unions watched with envy. In the United States the Air Line Pilots' Association (ALPA) working from its Washington palace, had successively increased the salaries and leisure time of pilots as their planes got faster and bigger until by 1975 the pilots' average flying-time was only fifty hours per month, and they could insist on having a crew of three in the cockpit. They effectively exported their union power through the pilots' international federation IFALPA and foreign pilots were quick to follow their example .... By 1983 fifteen percent of the 34,000 members of ALPA were on furlough, many with little prospect of being re-employed since "job-hopping" is rare among pilots; while in the meantime new airlines could often attract younger pilots for half the salaries'.

Commercial aviation is a notoriously cyclical enterprise. Pilots' fortunes wax and wane with those of the industry. Since the passing of the Airline Deregulation Act the industry has been through several cycles of expansion and relative stagnation. The terrorist attacks of September 11th 2001 caused a major perturbation with several airlines (like venerable carriers Sabena and Swissair) going out of business.

Such was the parlous state of the US industry following the attacks, that the US government intervened. It gave large sums of money to several US carriers. It would seem that even in the US, the country that spawned the airline deregulation movement, the interventionist instinct has not been entirely killed off. Morrison and Winston (1995, p. 162) comment: '[P]eople's allegiance is forever vacillating between the free market and government regulation. When an industry is struggling, they say the government should help. When a government programme falters, they call for market competition'.[28] What has changed irrevocably, however, is the structure of the industry. It is unrecognisable from the industry that carriers like Trans World Airlines, Braniff, Pan Am, Sabena, Swissair and BOAC/BEA, and institutions like the CAB helped to shape.

Since 1978 the industry has seen a 'slow burn' revolution that has impacted every single carrier and every single pilot. Of course, the venerable flag carriers still figure in the global aviation tableau. As Williams (2002, p. 172) states: 'National markets continue to be dominated by national carriers ... '. But as Read (2005, pp. 40–43) explains, national carriers are under considerable pressure from new entrants: 'At the time of writing [2005] Italy's flag carrier Alitalia is in serious financial difficulties but has been bailed out by a 400m [Euros] loan from the Italian government .... Several other state carriers have also announced recent losses, including SAS in

28 The US Government's interventions drew criticism from Europe.

Scandinavia, Austrian Airlines, Lietuvos Avialinijos in Lithuania, Olympic Airlines in Greece and Malev in Hungary'. To survive, says Read: 'Many national carriers have implemented tough cost-cutting measures cutting back both staff and services. This move has not gone down well with employees and several carriers have been plagued with industrial disputes as a result – including Aer Lingus, Alitalia, Olympic Air Lines and British Airways'. Read concludes: 'Despite efforts by countries, such as Italy, to keep their state airlines intact, the traditional national mould has started to break'.

New entrants like Jet Blue in the US and Ryanair and easyJet in Europe have played a major part in breaking the mould. In 2005, easyJet, based at Luton Airport just north of London, took delivery of its one hundredth aircraft. Ryanair, a low-cost carrier operating on an Irish Air Operator's Certificate (AOC), has helped create the new classless, mass-travel market.[29] As Creaton (2005, p. 255) explains: 'In the twenty years between 1960 and 1980 the number of people travelling between Ireland and Britain each year had remained static at about 800,000 .... Ryanair's first fifteen-seater aircraft took off from Waterford in 1985 .... Almost two decades on Ryanair is on course to carry 34 million passengers in 2005 .... The low-fares phenomenon championed by Ryanair has revolutionised air travel. One travel agent remarked that if you had advertised "cheap" airfares ten or fifteen years ago the public would have shied away from the offer for fear that the plane would fall out of the sky. "Now you can't sell a seat unless it's cheap", he [the Agent] said'.

Besides creating a new market the LCCs have begun to change the culture of commercial aviation, with Ryanair's Michael O'Leary in the vanguard. O'Leary's bluff manner, disdain for 'political correctness' and frequent use of expletives has set a different standard for aviation industry leaders. As Johnson (2004) notes in his *Times Online* portrait of O'Leary: '[H]e is gloriously inappropriate. But that's O'Leary for you – the man who redesigned the woman in the Ryanair logo to accentuate her charms. "She looked like a bloke with wings. Somebody said we should give her bigger boobs [breasts]. So we did. Some quango said we were demeaning women.

29 The further differentiation of commercial aviation's product range resonates with the post-Fordist dynamic of contemporary society where producers address niche markets with niche products (like no-frills air travel or business-only services – see Field 2006, p. 36). Murray (1989, p. 43) observes: 'Market niching has become the slogan .... Market researchers ... analyse "lifestyles", correlating consumption patterns across commodities, from food to clothing, and health to holidays'. To the extent that it elaborates its product range today's commercial aviation industry is post-Fordist. This may not be sustainable, however. As more and more airlines rationalise their product range to keep pace with dynamic low-cost carriers the number of products will reduce, leaving consumers once again with only a limited choice. It is quite possible, therefore, that as other markets embrace post-Fordism (product differentiation), commercial aviation will move in the opposite direction. In the not-too-distant future it may find itself selling a mass-produced, more-or-less standardised product to a mass-market. Re-massification would show commercial aviation in sharp relief against the broader trend. It would become an economic and marketing anachronism. (If this did happen, where would that leave such fashionable theories of de-massification as post-Fordism?)

F*** off. She's got bigger boobs [the premise being that happiness is correlated to breast size]. And the story got two half-pages in *The Sun*, worth £25,000 each"'. Carayol (2004) opines: 'O'Leary is ... innovative, fast, aggressive and unapologetic. He needs ... no cajoling to provide a stream of invective on just about anything that appears to stand in his way. The obvious targets are his competitors, and he singles out BA and Aer Lingus for special treatment ...'.

O'Leary is also wont to criticise his fellow no-frills travellers. Speaking in June 2004 he predicted a 'bloodbath' in the low-cost sector: 'We believe some of these [no-frills] loss-makers will go bust – some over this winter, some over the next two years. It may be BmiBaby or FlyBe in the UK – they're not all going to survive. They can't all sustain this level of losses for too long' (O'Leary cited in Clark 2004b). Despite the Ryanair CEO's 2004 prediction, as of November 2005 both airlines were still operating. Indeed FlyBe seemed to be thriving. *Airliner World* (2005b, p. 8) noted: 'The airline has developed a very successful business formula, using smaller capacity to provide valuable low-cost links between UK regions and into Continental Europe .... Following Ryanair's recent decision to scale back its services from London to Newquay by 12 flights a week, FlyBe has expressed an interest in operating from the South West airport'.

O'Leary has little time for politicians, too: 'Any time the politicians get involved in an industry or regulating an industry, they f**k it up. It's what they do best. If they stay the hell out of it, the industry works much better on its own by letting people compete. We had politicians and regulators running the airline industry for 50 years from 1945 to 1995, and they made a complete dog's dinner of it in Europe; we had the highest fares and the most restrictive practices. We've had 10 years of deregulation and it's transformed the cost of air travel around Europe' (O'Leary cited in Ashcroft 2004). Concerned about the impact of the EU's passenger compensation rules on the low-cost sector,[30] O'Leary (cited in Osborne, 2005) suggested the following course of action: 'I think we should blow the place up and shoot all the regulators ... '. 'The place' was the European Union's Brussels headquarters. Clearly the culture of commercial aviation has changed. The 'cosiness' has evaporated. The 'old school tie' mentality is under threat. There is less deference – indeed some airline executives appear to be unfamiliar with the term.

According to *Bloomberg News* (cited in Read 2005, p. 41) '... discount carriers now account for 12% of the ... flights operated annually worldwide, a percentage that has doubled since 9/11'. In 2005 there were 56 low-cost carriers in Europe alone (although it should be noted that some low-cost start-ups expired rather quickly).[31]

---

30 Even though low-cost carriers' ticket prices are (generally) lower than those of legacy carriers like British Airways and Air France, low-cost airlines have to pay the same level of compensation when things go wrong.

31 One start-up, the Irish-registered charter operator JetGreen, operated from 4 to 12 May, 2004. O'Leary has stated: 'The airline industry is too easy to enter. The barriers to entry are very low. The barriers to going bust are very low as well, but there's always somebody who's setting up an airline somewhere' (O'Leary cited in Ashcroft 2004).

The new entrants have brought with them a new vision and *modus operandi*. Ryanair, for example, does not recognise any pilots' union for the purpose of wage bargaining. At Ryanair wage bargaining is left to, as Hancock (2004a) puts it: '[S]elf-appointed worker representative groups in each country where the firm has bases'.[32] Clearly such changes have had a major impact on pilots, as the following accounts demonstrate.

*Pilots' lived experience of the line*[33]

In his book *Nine Minutes, Twenty Seconds* Gary Pomerantz (2001) describes the crash in 1995 of a US commuter aircraft (a twin turboprop, three-crew Embraer EMB-120 Brasilia) operated by regional carrier Atlantic Southeast Airlines (ASA).[34] As can be seen from Pomerantz's sketch, life for the crew was far from glamorous: 'The crew was coming off a hard weekend of flying: heavy rains, interminable ground delays, surly passengers. On Saturday night, as midnight approached, Gannaway, Warmerdam, and Fech had struck a forlorn pose: standing at the curb in the lamp-lit rain at the Tri-Cities Airport in east Tennessee, the end of a gruelling fifteen-hour day that began in one small southern city and ended in another, standing together but feeling alone and waiting for a motel van that was, like everything else that day, late. When they arrived at the motel, Gannaway made a quick call home to his wife .... The captain had left his house at six A.M. and seventeen hours later his voice was heavy with fatigue. "Hard day", he said. "Bad weather". The motel's night manager unlocked the motel kitchen for Fech. She took a small box of cereal, a carton of milk, a Styrofoam bowl and plastic spoon to her room. Not exactly five-star dining, but it helped her sleep. The crew made it back to the Tri-Cities Airport by seven-thirty the next morning and returned home to Macon in time for Sunday dinner. Now, up and ready on a Monday morning, they had six flights scheduled, with an overnight in Albany, Georgia. More than a living, this was their life' (Pomerantz 2001, pp. 13–14). While their working life could be hard, pilots who flew for regionals were often paid less than those who flew for the US majors: 'A veteran ASA captain ...

---

32 According to Michael O'Leary, Ryanair's pilots are some of the best paid in Europe (Hancock 2004b). Cottell (2004) notes: '[Ryanair] says that its average salary of £50,582 is among the highest of any airline in Europe, and compares favourably with British Airways (£37,602) and easyJet (£41,384)'.

33 'The line' is the term given to the sharp end of commercial aviation – actual point-to-point revenue-earning operations.

34 At the time of the accident, ASA operated 83 commuter aircraft. Because of their limited capacity they were operated to less stringent Federal Aviation Administration standards than those applied to the large-capacity aircraft (Boeing 727s, 737s, 757s, etc.) operated by the majors. According to Pomerantz (2001, p. 29): '[C]ommuter pilots flew longer hours and received less training than pilots at the major airlines (of commuter pilots surveyed by the NTSB [National Transportation Safety Board] 87 percent said they had flown while fatigued due to inadequate rest periods)'.

in 1995 topped out at about $60,000 a year, while a Delta[35] captain made upward of $180,000, with a superior retirement package. The financial pinch was especially severe for ASA first officers. Their starting pay was about $15,000' (Pomerantz 2001, p. 31).

Given commercial aviation's symbiotic relationship with the world at large, major upsets like the two Gulf Wars, terrorist attacks of September 11th 2001 and the 2005 Asian Tsunami have major impacts on the industry and on the career prospects of those within it.[36] Newly qualified pilots may find it difficult to find work. Employed pilots may be furloughed (a favourite tactic at Pan Am). Others may find themselves moving from one short-term contract to another.

Despite this vulnerability the industry maintains an optimistic persona. In the book *Century of Flight*, published to commemorate the first 100 years of aviation, one of the contributors remarked: 'Perhaps it is the money that attracts people to the job, maybe it's a status thing, or maybe it's just a love of flying and the opportunity to travel the world .... There are an increasing number of openings for commercial pilots ... ' (Bridgewater 2003, p. 178). More recently the General Secretary of the British Air Line Pilots' Association (BALPA) remarked: 'Growth in air travel over the next decade, and the retirement of longer serving pilots, makes it a good time to consider becoming a Pilot – for those who are determined' (McAuslan cited in BALPA 2005a, p. 2).

Aspiring pilots certainly need to show determination. Although the industry enjoys steady, long-term growth, career prospects for pilots fluctuate. As one aspiring First Officer explained in an article for BALPA's house magazine *The Log*, getting that first job can prove very difficult: 'CVs more often than not don't get a reply and to some extent this is understandable. I spoke to a Chief Pilot recently ... who told me that he had received over 2,000 CVs in a three month period .... Nearly all the pilots I come into contact with can relate to the difficulties in getting the first job .... With so many people available, the airlines, especially the smaller ones, don't need to advertise to fill a handful of vacancies .... [T]he decision to pursue a career as an airline pilot is one that involves a lot of hard work, huge expense and great sacrifice with no guarantees of a job on receipt of your licence' (Reece 2004, p. 13). The author of this article had put himself through an Integrated training course at a major UK-based air training school. The training was costly[37] and taxing: 'Fortunately, I had enough equity on my property to pay for the training without having to take out a loan of any kind. Taking the Integrated route ... meant carefully budgeting for food, accommodation, bills and general living costs for 18 months which is a long time without any form of income .... Lectures started at 8.30am and finished at 4.30pm each day and this was followed by at least 4 hours of study at home each evening ....

35 Delta Airlines was a US major.

36 There were fears that the July 2005 terrorist attacks on London would impact tourism and therefore the airlines.

37 In 2005, BALPA estimated the cost of an Integrated course at between £50,000 and £60,000 (BALPA 2005a, p. 5).

The volume of work is really quite extensive and the course certainly has its stressful moments ... ' (Reece 2004, p. 12).

Some 15 years ago, the British Broadcasting Corporation made a documentary series about British Airways and its staff. A book was published to accompany the series. British Airways' recruitment drive and training regime were described as follows: 'When British Airways placed two advertisements for new-entrant trainee pilots, one in *The Daily Telegraph*, the other in the monthly aviation magazine, *Flight International*, it received 6,800 replies .... Cadets spend seventy weeks in training at Prestwick, and it is no holiday camp .... The course runs a five-day week, although some Saturday work is also included; Sundays are free. Throughout the seventy weeks just two holidays, totalling three weeks, are scheduled .... Failure to keep pace with the demanding and ever-intensifying curriculum, both on the ground and in the air, could result in a request to pack bags and leave' (Reed 1990, pp. 110–113). It is clear that would-be First Officers are prepared to subject themselves to significant hardships to gain their commercial licence. It is also clear that there are significant risks attached to this process. These include the risk of failure to complete the course and the risk that, having gained one's licence, one will not be able to secure a position.

The British Air Line Pilots' Association (2005a, p. 6) has highlighted the volatility of the industry and its impact on pilots' career prospects: 'Your ... chances of being hired ... are entirely down to supply and demand. You can be the greatest pilot in the world, but if there are no vacancies, it may take some time to obtain employment as a Pilot .... It is very common for employers in this business to be convinced that they need no one extra on Friday only to find that the whole situation has changed by Monday. If you are in the right place at the right time ... the job could be yours'. The use of the phrase 'right place at the right time' is noteworthy. It draws our attention to an important difference between the world of aviation training and the world of aviation employment. While the former is structured and predictable, with syllabuses to be followed and exams to be passed, the latter is uncertain and volatile. If there are no openings for low-hours pilots then a great deal of effort and financial outlay may be wasted. Commercial flying is hardly a risk-free career. As one 32-year-old First Officer has noted: 'I left school at 18 and became an insurance broker. Then moved across to computer system training and office management. Eventually, *I gave up that nice steady career to become ... an airline pilot*' (my emphasis) (McLaurin 2005). Glossy descriptions like that offered by Bridgewater (see above) can be misleading.

As indicated by Pomerantz's description of the 'Tri-Cities Crew', flying can be unglamorous, exhausting and lonely. It is interesting to contrast Pomerantz's vignette with Chuck Kraft's impressions. Pomerantz's Tri-Cities Crew lived in a world of stopovers at tertiary airports on minimum rest; of late meals ate in front of the TV in some nondescript, freeway-hugging Motel; of grabbing a few hours sleep in a well-used bed with a plastic-covered mattress. Kraft glimpsed a world of measured relaxation in comfortable hotels in glamorous locations like Paris and Hong Kong.

Pomerantz's Tri-Cities Crew is an allegory for the modern flying experience, as may be gathered from the following testimonies:

*UK-based First Officer (female)*

> There are two bad things [about commercial flying]. Firstly the hours. Starting work at midnight on a cold, wet January night can really test your enthusiasm. Secondly, airline food. Most people get to eat it perhaps twice a year – we get it every day! (McLaurin 2005).

*US-based First Officer (male)*

> I think that 'airline pilot' is a likely candidate for *number 1 misunderstood job*. Most of the public thinks they have an idea what we do, and very few actually have a clue .... [T]he average person is widely missing the mark by thinking pilots are overpaid, underworked playboys frolicking with the 'stewardesses' (that's Neanderthal for *flight attendant*) on the Hawaii layovers when they're not home with the missus for 30 days a month .... If anyone in the public still thinks of this as a glamorous industry, their head is still in the 60s. Airline life, while still enjoyable to some, is hard. You're spending hundreds of hours each month away from home, and unless you're senior at a major airline, that time is more likely to be spent at Billings or Detroit than Hawaii or Europe. You're not staying at 4-star hotels ... most crew hotels are simply accessible and functional .... International pilots – the rather senior ones – may work 10 days a month .... The rest of us are working far more. I'm currently getting 12 days off in a 35 day bid .... [A] 4-day trip for me [might see] overnights in Sun Valley, Edmonton and Calgary .... [W]hen you return late at night, four days later ... [y]ou're usually totally fried [i.e. exhausted] .... Anybody looking to enter such a turbulent, troubled industry needs to take a long, realistic look at what the job is like before they expend significant time and money to pursue this career ' (cited in Blogging at FL250 2005).

*UK-based First Officer (male)*

> Working as a pilot is exciting, but of course there are downsides to the job. It involves getting up early, working weekends and during holiday periods, and being 'on call' at certain times. It is important to realise that it is hard work and that you will be required to work when the airline needs you and not when you might prefer to work. If you are thinking of going into this profession I would really recommend trying to talk to an airline pilot about their job, particularly on an informal basis and over an extended period, so you get a 'real' understanding of what the job is like ... (cited in Prospects.ac.uk 2005).

Commercial flying, therefore, is often unglamorous. It also has the potential to cause stress and fatigue. While stress and fatigue are difficult concepts to define[38] there is

---

38 Stone and Babcock (1988, p. 536) note: 'Fatigue and its associated hazard, performance decrement, have fascinated researchers for many years, but the elusive character of fatigue has escaped successful scientific description'.

evidence of its existence amongst flight crews[39] (although the evidence presented here consists only of *subjective* assessments). Consider, for example, the following statements made by the Captains and First Officers of one UK-registered low-cost carrier:

> [I feel] very tired. Certainly as the fifth sector was winding down it was harder to concentrate. All those things that you take for granted you were actually concentrating more on because you are thinking: 'I may be missing things here'. Certainly my reactions were much slower. I could feel myself being less sharp, shall we say .... (This Captain had just completed a five sector day shuttling between Stansted Airport and Edinburgh.)

> I have been with [this airline] now for seven months, and last winter I had a choice of three jobs to go to, and two of them were effectively sweat shops. However the ... lifestyle [at this airline] was sold to me [on the basis] that we only do three or four sector days .... You can plump for night stops or you can stay away from night stops .... When you have got your days off you finish on an early and you start on a late. They really look after the crews. Therefore this is the place to be .... I have got to say since I have been here that is not actually the case. Now obviously I understand there is a transitional period in [this company] ... it is expanding rapidly and they are trying to make the airline ... look as profitable and forward planning and 'go getting' to make sure that the investment comes in .... But now that they have actually achieved that ... there are people obviously in high places who have looked at it and said: 'Look we can actually make the crews do this and we are actually profitable by doing this and yes, people are whinging and there are some people leaving, but on the whole it is achievable, therefore we might as well keep this work-rate up'. (This reflection on the company's push to maximise turnover and yield during the Summer of 2001 was made by the same Captain who had done the five-sector Stansted/Edinburgh duty.)

> [I felt] tired [this morning] .... More tired than I normally am on an early start, but then it is the end of the week .... A number of earlies ... and long days .... I'll be jaded tomorrow ... and I'll be fairly bright on Monday ... I'll be back to work on Tuesday ... this has been a hard week ... six days on ... an extra sector .... I think I've hit my limit ... I've hit my limit now .... A good night's sleep would get me back into being safe. (This Captain was interviewed on a Friday night after completing his sixth consecutive day of flying. He was scheduled for three days off (Saturday, Sunday and Monday).)

> I said I am not tired this morning, but I think this is very unusual. Usually I am very tired. Today was OK because I try hard to go to bed early and I went to sleep a little bit yesterday afternoon. Honestly, I keep sleeping to try to catch [up] with this duty time .... [I]f I want to see somebody in the evening ... if I want to go for dinner ... I can't .... If you are six days on duty you spend your six days ... working ... and you try to adapt to this pattern ... and next week I am going maybe late [late-start duties], and so I will have to re-adapt to

39 There is concern that fatigue is also afflicting cabin crew. In 2004 a spokesperson for the Association of Flight Attendants remarked: 'We're getting a lot more complaints about fatigue than last year, as the industry is trying to get more productive' (Kolander cited in *Air Safety Week* 2004b, p. 1).

another pattern .... (This First Officer (whose first language was French) was trying to get a command at the subject airline.)

I really do hope [the rostering] does improve. I feel there are ... not really loopholes as such ... but there are opportunities to exploit the way the rules are written at the moment .... I am flying a plane sometimes absolutely exhausted .... Even if I was fully familiar with what I was doing ... fully on-line for about a year or two ... I would still find it very tiring .... (This First Officer suggested that although the airline never broke the rules relating to flight time limitations, in his opinion it worked its flight crew as hard as it could.)

[A]t the end of the day there is that commercial pressure to keep the operation going .... Yes, that is what it is all about. It is all about turnover, basically. I guess there are subtle, sort of, inferences there, isn't there, for fatigue? (This observation was made by a Captain.[40])

It would be wrong to imply that this airline's behaviour was markedly different from that of any other low-cost carrier. While the low-cost model is subject to variation,[41] its basic tenets tend to produce *broadly similar* forms of behaviour (and therefore culture and practice) the world over. Holloway (2002, p. 95) describes the model thus: '[Low-cost carriers must] build a cost structure which ... can ride-out the fare wars ... and ... can sustain low prices. To achieve this low-cost/low-fare carriers require ... *high asset and workforce productivity*: [Low-cost carriers must] ensure the culture of the airline supports a low-cost/low-fare strategy (*an essential ingredient being flexibility and a willingness to "pitch-in"*)' (my emphasis). Ryanair's Michael O'Leary (cited in Arlidge 2001) has commented: 'Everything is geared towards efficiency .... We are ... more aggressive than our competitors'. In 2004 he said: 'Airlines are very hard to run and if they're badly run they're very hard to turn around .... You've got to be cleverer every year to keep lowering costs' (O'Leary cited in Ashcroft 2004).

---

40 These statements from Captains and First Officers were first published in an academic journal paper entitled 'Flight crew stress and fatigue in low-cost commercial air operations – an appraisal'. See Bibliography for details (specifically Bennett, S.A.). The journal paper reproduces one interviewee's roster as an Appendix.

41 The model applied by easyJet, for example, varies in certain respects from that applied by Southwest Airlines and Ryanair. While Southwest and Ryanair operate a single aircraft type (the Boeing 737), easyJet operates two types (the Boeing 737 and Airbus A319). JetBlue's use of aircraft from two manufacturers (Embraer and Airbus Industrie), investment in satellite-based information and entertainment systems and provision of a small number of complementary snacks is an obvious variation on the LCC theme (that preaches standardisation and cost-cutting (Nicholls 2006, pp. 26–30)). Kuwait's Jazeera Airways is also varying the theme: 'For a surcharge passengers can upgrade to Jazeera Plus and benefit from separate check-in desks ... a greater baggage allowance, free in-flight service and more space ... ' (Maslen 2006, p. 50). In 2005, Indian low-cost airline Kingfisher announced plans to operate an Airbus A380. The A380 is not thought of as being a classic low-cost vehicle (unlike the B737 or A319).

As mentioned earlier, the dynamism of the low-cost business model (and of those who apply it) has generated enormous change in the global aviation industry. Maslen (2003, p. 123) comments: 'The low-cost segment is now one of the most fluid markets, having seen tremendous growth, particularly in the early years of the new century .... [S]ince its launch in 1970, Southwest has expanded its fleet to over 350 aircraft, representing an average 17.5% capacity increase per annum. With airlines like Ryanair now even offering free flights, the low-cost sector really can be said to have opened up air travel to the mass market'.

North American carrier Southwest Airlines pioneered the low-cost model – a model the airline implements with zeal. As Palmason and Kramer (1999, p. 21)[42] explain: 'Southwest achieves ... low costs through very high aircraft, equipment and personnel productivity. Aircraft achieve fleet-wide utilisation in excess of eleven hours flight time and nine flights per day .... Most aircraft ... turnarounds are scheduled for 20 minutes at the gate, supporting the high equipment and personnel utilisation'. According to Palmason and Kramer (1999, pp. 22–23) industrial relations at Southwest have shown a marked improvement over the years: 'Southwest Airlines has excellent management that considers people to be a crucial resource .... The conflict of the mid-eighties helped forge a trust based on mutual respect between our union and our management .... The folks who run this airline did not build America's lowest cost operator by giving away money. The excellent relationship between the union and management does not prevent disputes, it merely allows us to work through them in a timely fashion without warfare'.

Clearly, then, pilots' experience of low-cost operators varies. Some pilot groups seem relatively happy, while others seem less so. Much has been made of Ryanair's reluctance to recognise pilots' unions for the purpose of wage bargaining. There is suspicion that Ryanair has something to hide. In 2002 a reporter for *The Observer* canvassed the opinion of some of Ryanair's Stansted-based pilots. She reported: 'Pilots earn between £50,000 and £125,000. Serving pilots hanging out off duty in the bars of Bishop's Stortford last week ... said that working up to 900 hours a year was not a problem and they enjoyed their "work-hard, play-hard lifestyle". One, a cheery, 30-year-old blond eastern European ... said: "Ryanair puts safety as a top priority and we never disobey air traffic control. I would work up to 1,000 hours if I could. We have rests, it's not a problem. I'm really happy and as they are expanding fast I'm likely to get promoted quite quickly from first officer to captain"' (Walters 2002).

As to union recognition, Ryanair states that the majority of its pilots do not want it. Speaking in 2004 a Ryanair spokesman commented: 'There is no evidence to suggest Ryanair pilots want the trade unions to represent them. The last time this matter was put to a secret ballot, conducted independently in October 2001 only 18% of Ryanair's BALPA members supported trade union recognition. The reason

42 In 1999 Captains John Kramer and Steven Palmason were members of the Southwest Airlines Pilots' Association.

the majority do not want union recognition is because they achieve better pay and bigger pay increases by dealing directly with the company' (cited in Cottell 2004).

Of course, there are often two sides to every story. In 2004 some of Ryanair's Dublin-based pilots took issue with the company's plans to introduce new Boeing 737 marques. Ryanair wanted to tie its retrained pilots to the company for five years. One pilot remarked: 'My colleagues are fed up with the erosion in our terms and conditions' (cited in Hancock 2004b). In the same year Ryanair, faced with a trade union drive to boost membership, issued the following communication to its Stansted-based pilots: 'We have no objection to any Ryanair pilot joining BALPA, the Taliban [Afghanistan's deposed Islamic Fundamentalist rulers],[43] the Monster Raving Loonies [a group that sometimes puts up candidates at UK General Elections] or indeed the Moonies [a sect]. Each individual is perfectly free to join whatever organisation he/she chooses. If you want to waste £1,000 we recommend fast women, slow horses or even greyhound racing. At least you'll have a few minutes of fun' (cited in Cottell 2004).

During the exchange, the trade unions accused Ryanair's management of, amongst other things, penny-pinching. As Cottell (2004) explained at the time: 'The unions argue that an aggressive pursuit of profits pervades the company's management. This, they say, is reflected in the way staff are forced to pay for their uniforms and airport car parking, a refusal to pay trainees and a hostile anti-union culture'. In 2002 *The Observer* newspaper reported one Ryanair pilot as saying: 'The culture at that company is one of pressuring pilots to work very long hours and it has only got worse since I left' (cited in Walters 2002). In 2004 O'Leary commented: 'The only pressure that pilots are under is to prioritise safety' (O'Leary cited in Johnson 2004).

It is important to note that worries about rosters and other operational parameters are not entirely new. In 1969, David Beaty published *The Human Factor in Aircraft Accidents*, one of the first explorations of the human factors aspects of commercial aviation. *The Human Factor in Aircraft Accidents* was, in all but name, a sociology of commercial flight crew. Reading the book today it is clear that even in the pre-oil-shock 1960s, commercial aviation had its problems. Then, as now, it lurched between expansion and stagnation. As Beaty (1969, p. 18) noted: 'Redundancy often appears to be just round the corner. Airline statisticians have rarely been able to get their predictions right on how many pilots will be needed in the years ahead [this observation resonates with that made by BALPA (2005a, p. 6) some years later (see above)]. An airline's share of passenger traffic fluctuates. Companies suddenly get scared of having a great many empty seats and do not pursue an even, long term policy. As a result there is either a world shortage of pilots or there are far too many of them'.

---

43 Given the Taliban's appalling record on human rights, destruction of a UN-designated World Heritage Sight in 2001, systematic misogyny (including the denial of educational opportunities to women), and quiescence in religiously-motivated terrorism (Afghanistan's terrorism training camps), this reference would seem a tad tasteless.

Chuck Kraft's encounters with well-fed, bronzed and urbane long-haul flight crews in glamorous locales created such a favourable impression that he resolved to leave the military and train as a commercial pilot. If Beaty is correct, however, even in the expansive, Swinging Sixties the 'lived reality' of the long-haul pilot fell some way short of Kraft's ideal-type relaxed and well-adjusted globetrotter.

As this extract from *The Human Factor in Aircraft Accidents* shows, many of the problems that confront pilots today were prevalent four decades ago: 'Along the route, sleeping alone in strange beds at varying temperatures, after strange food in different latitudes, no wonder the pilot finds difficulty resting. Officially on this point there is sympathy. Quiet rooms in the best hotels are provided – but the hotels naturally remain geared to local time eating and sleeping habits. Coming in from a long trip, particularly if there has been a difficult landing, a pilot will be "wound up", or in psychological terms "aroused". He does not feel the need to sleep until he has "unwound". Asked by a pilot how he could be sure of some rest *en route*, an airline doctor's prescription was, "Drink beer, take drugs or sleep with the stewardess" .... [H]e tends to return [home] not so much in the role of head of the family, but as a mixture of three people – a rich uncle loaded with presents, an irritable authoritarian grumbling about their lack of discipline, and a tired invalid for whom the rest of the household has to keep quiet and to whose haywire sleeping and eating habits the whole house has suddenly to become geared' (Beaty 1969, pp. 20–21).

Beaty also explored issues of general health, stress and fatigue through the medium of a questionnaire survey of pilots' wives conducted in 1965. The responses are illuminating. They might have been made by the partners of today's pilots:

> He has too short a time between trips and doesn't have time to get rested.
>
> He comes home from a trip completely tired out. The next day is lost to fatigue.
>
> He is working for a physical breakdown, real serious one.
>
> He says he won't have to worry about retirement because he will never make it.
>
> It has made him an old man.

Beaty included some comments pertaining to flight time and sector pay:

> This 85 hours per month is for the birds [its a myth].
>
> [T]he way the contract stands now, you can hardly blame the poor jerks for selecting to fly when they should be home in bed. Few people, pilots included, can afford to kiss off over a hundred dollars by missing a flight because they have a runny nose.

Beaty also reproduced comments made about the integrity of pilots' marriages:

> No sex life since flying jets.

His fatigue has ... altered the frequency of our sex-life (cited in Beaty 1969, pp. 23–24).

It would appear, then, that issues associated with the economic state of the industry, the impact on circadian rhythm of intensive intercontinental flying, drug abuse (i.e. alcohol abuse), interruptions to normal family life, stress and fatigue have been around for a long time. So, too, have concerns about automation and its impact on the professional standing and public image of flight crew, especially that of Captains. Beaty's (1969, p. 17) observations are insightful: 'An airline captain is ... conscious that ... his status has changed .... [W]ithin the company his status has shrunk. On the aeroplane he remains master. But he has lost much of his personal contact with the passengers that he used to have simply because often there is not the time to go back and talk to them [today, of course, locked cockpit doors and the banning of visits to the flight deck mean that there is even less contact. See, for example, Bennett (2002a)]. His crew is now smaller than on piston-engined aircraft – the navigator and radio officer have both gone, and on some aircraft so has the engineer'. It is informative to contrast Beaty's description with that offered by Quinn (2003) of the Captains of BOAC's C-class Empire flying boats or early turbojet-powered aircraft like the De Havilland Comet. (One of Quinn's interviewees described them as 'lords of the air'.)

*Pilots' experience of automation*

The concerns about automation noted by Beaty in the 1960s are prevalent today. Indeed, with the advent of innovations like satellite navigation, powerful, performance-optimising flight management computers (FMCs), glass-cockpits with multi-function displays (MFDs), fly-by-wire controls, two-person flight decks (even on aircraft as large as the Boeing 747-400 and Airbus A380), automatic data uplinks from ground stations, autothrottles (i.e. full authority direct engine control (FADEC) systems) and advanced, multi-mode autopilots, such concerns may have intensified.

Of course innovations such as these do not simply appear as if by accident. As those celebrated critics of technological determinism, Wiebe Bijker and John Law (1992, p. 3), explain, there is no *given* technological trajectory. Technologies are the product of premeditated, purposeful artefaction: 'The idea of a "pure" technology is nonsense .... Technologies do not ... evolve under the impetus of some ... inner technological or scientific logic. They are not possessed of an inherent momentum. If they evolve or change, it is because they have been pressed into that shape'. As artefacts – the product of human intent and labour – all technologies have an underlying logic or politics ... *a purpose beyond themselves*.

In commercial aviation that logic is very often concerned with cutting costs. More competition creates more pressure to cut costs. This pressure transmits to aircraft manufacturers and legislators where it influences technological and regulatory innovation. Market forces impact every aspect of commercial aviation, from the design of aircraft, powerplant and avionics, to recruitment and career prospects, to national and international regulation. As BALPA's Director of Strategy has noted: 'Arguably ... the biggest challenge [facing us] is economic. The industry has changed and changed utterly .... [S]ometimes economics, regulatory issues and business strategy seem far from the everyday concerns of the flight crew community. It's understandable to focus on your day-to-day problems .... However, it is important that we ... concern ourselves with the wider issues which shape our industry' (McGurk 2004, p. 4). According to Koonce (1999, p. 12) economic considerations also shape airspace management policies and technologies: 'Economics are pushing a restructuring of the way in which the airspace system works [with] [c]oncepts such as data links between controlling agencies and the aircraft they control, free flight to optimise flight efficiency, comfort and safety [and] automation of weather observation and dissemination ...'.

In his 1997 book *Aviation Automation: The Search for a Human-Centred Approach* Billings identified seven levels of flight-deck automation. These ranged from the lowest level, 'Direct Manual Control', through 'Assisted Manual Control', 'Shared Control', 'Management by Delegation', 'Management by Consent' and 'Management by Exception', to the highest level of automation, 'Autonomous Operation'. Applying his classificatory scheme to the Airbus A320 (at the time a popular two-crew, two-engined short-haul airliner), Billings (1997, p. 203) commented: 'Airbus [provides] *shared control* as the A320's basic control mode. Pilots' control inputs are considerably modified and shaped by the flight control computers; envelope limits prevent them from exceeding predetermined parameters [envelope protection][44] .... Under normal circumstances, the aircraft automation is responsible for much of the inner-loop control, although control laws are tailored to respond in ways that seem natural to the pilot'. Billings (1997, p. 203) comments that as a component of this form of control architecture: 'The pilot remains in the control loop, although it is an intermediate rather than the inner loop'.

Feasibly, says Billings, the pilot could be squeezed out altogether: 'In the near future [the aircraft] will communicate with ATC computers, accept and execute ATC clearances, and report its location when not under radar coverage .... Some have noted that automation usually performs all of these functions correctly, that it does not become tired or distracted or bored or irritable, that it often speaks more clearly and succinctly than pilots do, that its data stream will be easily comprehended by ATC computers in any nation, and that it does all these things without complaints .... [S]ome air carriers have mandated that [automation] be used wherever possible' (Billings 1997, p. 201). Regarding Billings' last point, Buck (1994, p. 183) notes: 'The modern airplane is flown mostly on autopilot, and airlines encourage – in some cases demand – that pilots use the autopilot for all flying after takeoff through and including landing. At least they did in the first flush of the new airplanes, but this

44 Billings (1997, p. 209) defines envelope protection or limitation as follows: 'Certain parameters (bank angle, pitch, or angle of attack) cannot be exceeded by the pilot except by turning off portions of the flight control computer systems or flying outside their cut-off values .... Predetermined thrust parameters also cannot be exceeded'.

has been modified to some extent after experience with the airplanes'. Buck's last point alludes to one of the key drivers of the Modernist project – 'technological exuberance'. Technological exuberance may be defined as the obsessive pursuit of automation to the exclusion of all other considerations (like worker morale, cost, economic efficiency and, possibly, safety). The condition can be thought of as a 'zealotry of mechanicism'.

While automation has reduced pilots' physical interaction with the aircraft and its systems, it has increased their cognitive workload. Koonce (1999, p. 11) asserts: 'With highly automated systems and reliable equipment the flight crew physical workload has been significantly reduced; however, there seems to have been a resultant increase in the cognitive workload'. Given that human operators are not especially good at passive monitoring for extended periods, situation awareness (SA) has become a major issue for systems designers. As Koonce (1999, p. 12) explains: '[Modern] navigational systems are extremely precise, and they are integrated with the autopilot systems for the capability of fully automated flight from just after takeoff to after the airplane touches down on the runway. The pilot is becoming a passive observer of the airplane's systems while the automation does the flying. A challenge for the designers is what to do with the pilot during the highly automated flight .... Recently there has been a great amount of attention to the concept of situation awareness in the advanced airplanes .... Accidents have occurred in which the flight crew members were not aware of their location with respect to dangerous terrain or were unaware of the current status of the airplane's systems when that knowledge was essential for correct decision making'.

In 2004, BALPA published an article entitled 'From airline pilot to airline machine-minder' in its house journal *The Log*. The article set in motion a lively exchange of views in the journal's letters pages. The article's author made a number of claims, one of which was that error rates were, in part, a correlate of efforts to control error. Scott explained his counter-intuitive point in the following way: 'Airline managements are responsible for setting out standard operating procedures (SOPs) that require pilots to operate their aircraft in a prescribed way .... [M]odern pilots are severely constrained by literally thousands of flight crew orders (FCOs), policies, rules and regulations. Each of these rules restricts the decision-making ability of the pilot.[45] In this context it is notable that 80% of accidents are now ascribed to human error rather than the 20% of earlier years' (Scott 2004, p. 5).

Put another way, the author of 'From airline pilot to airline machine-minder' argued that (human) error rates are to some degree constructed by aviation's rules and regulations (in the same way that levels of crime are constructed by a country's system of laws). The more rules and regulations you have, the greater the opportunity

45 The close direction of a worker's *modus operandi* is known as Taylorism, after Frederick W. Taylor, the architect and chief apostle of scientific management. Smither (1988, p. 10) notes: 'Taylor ... believed that there was one best way to perform any particular job, and efficiency required that procedures for performance be standardised'. He also says: 'Many workers did not like being told how to accomplish their tasks ... '.

for/likelihood of transgression. The author concludes: 'As the[se] constraints ... have caused pilots to be seen as the major source of errors, so managers have reacted by constraining them further! For example the airline for whom I work has increasingly discouraged manual flying and the use of manual thrust control on the Airbus fleet. This has led to a de-skilled pilot force with a consequent rise in manual flying errors. The airline's response ... was to ban the use of manual thrust in normal operations .... Since Airbus sidestick inputs are mediated by flight control computers this means there is no longer any direct pilot control of the aircraft. This ban is likely to lead to further de-skilling and further increases in errors' (Scott 2004, p. 5).

According to Scott, the locus of control has shifted away from the Captain (and her/his First Officer) towards, as he puts it, 'automatics and managements'. He summarises: '[T]here has been a historical shift from fully human control towards fully automatic control of aircraft. Currently we have a hybrid system in which control and decision-making are shared in an ambiguous and poorly understood way between pilots and computers. With airline managements firmly committed to constraining pilots rather than supporting them we have reached the point of no return. The stick and throttle are history. The airline pilot era has ended; the airline machine-minder era has begun' (Scott 2004, p. 6).

Scott's pessimistic vision resonates with the views of academics like Braverman and Cooley. Braverman's determinedly Marxist view of the labour process held that, in Theobald's (1994, p. 111) words: '[I]n an era dominated by large corporations ... managers ... must constantly seek to redesign and modify the labour process in order to achieve maximum profit. To reach this goal management needs to assert as much control over the labour force as possible'. Frederick Taylor's scientific management is one means of regaining control of the labour process. Theobald (1994, pp. 111–112) describes the logic of scientific management thus: 'Wherever possible worker autonomy is restricted and jobs redesigned in order to transfer knowledge, discretion and control to management'.

Engineer and trade unionist Mike Cooley's work is also relevant to the question of who controls the production process. As Theobald (1994, p. 116) explains: '[Cooley argues that] specialised knowledge is being abstracted from individuals and embedded in organisations. [Cooley] sees the computer as the "Trojan Horse" through which Taylorism is being introduced into intellectual work. One view of computerisation ... is that it will free human beings from routine ... tasks, leaving them free to engage in more creative work. Cooley questions this assumption, maintaining that the same deskilling processes that have been eroding craft work ... are increasingly degrading mental labour'. As ever with academia there is a counter-view. Thus, the deskilling thesis advanced by Cooley, Braverman and others is challenged by academics like Dahrendorf and Kerr, who argue that as industrial societies develop there is *more* demand for skilled labour.

So who is right? Is labour being deskilled, or is the demand for skilled labour increasing? The answer may not be clear-cut. It may be that the locus of skills development, and repository of wisdom is simply migrating from front-line workers (like pilots) to designers, manufacturers and maintainers. As Theobald (1994, p.

112) observes: 'As technology grows more complex, more qualified designers, builders, engineers, maintenance and repair personnel are needed'. These upstream personnel have the job of 'hard-wiring' devices like FMCs, GPWS units and TCAS transceivers/processors with pilots' accumulated (and often tacit) knowledge and wisdom. The task of downstream personnel (pilots) is simply to read and action the manuals, SOPs and check lists produced for them by the industry's legions of desk-bound, computer-centric technocrats.

Not unexpectedly, pilots have different opinions about automation. Scott's views lie at the more pessimistic end of the spectrum of pilot opinion. Theobald's (1994, p. 114) discourse on worker resistance is useful when considering Scott's pessimistic vision, providing as it does a measure of reflexivity: 'It has been observed that professional and technical workers ... often become exceedingly preoccupied with their professional status precisely when it is under threat, usually through the imposition of increased ... control'.

Buck (1994, p. 183), who started his commercial flying career on Douglas DC-2s and ended it on Boeing 747s, is more sanguine: 'Because of the computers and programming, "wise" people like to say that the pilot has become a systems manager. This expression grates on my sensibilities, because whether you're pushing buttons or moving old-fashioned controls, you'd better have a lot of flying knowledge, background and judgement – which makes you a pilot, not a systems manager. The computer for all its miscellany is just another tool for the trade'. As far as Buck is concerned computerisation does not mean the permanent deskilling of pilots. Rather, pilots' skills and knowledges are simply duplicated to computer programmes. When those programmes are overwhelmed or outmanoeuvred (or when the silicon chips that run them are fused by spilled coffee or fry themselves) the knowledgeable and skilled pilot is able to intervene.

Bearing in mind Buck's more sanguine view of automation, it is interesting to note that Scott's article generated a good deal of support in the letters pages of *The Log*. One correspondent encouraged readers to ponder the possible consequences of some airlines' attempts to set stricter criteria for hand-flying: 'Aeroplanes still operate in a hostile environment, and are still subject (not as often now, admittedly) to systems failures .... Consider how you might feel if faced with a manually flown, 3 engined, localiser-only approach, at night with no forward visual reference due to sand-blasted windscreens. I saw it done, and done well; but by someone practised in hand flying' (Greaves 2004, p. 37).

Another correspondent talked about the robotisation of flight crew through the application of ever-more prescriptive Standard Operating Procedures (SOPs): '[I]f a management does succeed in browbeating its pilots into following every dot and comma of narrow, iron-walled procedures, is it any safer? I think not. Because now you have robots who are afraid to think, can't think, won't think, but will follow the rulebook slavishly, even in defiance of common sense and good airmanship ... ' (Jones 2004, p. 37).

A third correspondent contended that while the more extensive use of automation may reduce incidents in the short-term, it may serve to sow the seeds for an accident in the longer term. This correspondent also drew attention to the possible adverse effect on pilots' motivation of increased automation and more frequent application of Taylorism: 'It needs to be realised at the highest echelons of airline management that if they try to enforce a policy of "no manual flying unless absolutely necessary", they will not risk lowering the morale of the pilots, they will lose it altogether .... Periods when ... our motivation ... declines, are nearly always identified with a related drop in performance. However professional we like to think we are, human nature ... will always prevail: if we lose motivation we lose performance' (Matthews 2005, p. 37).

A fourth correspondent commented: 'I am becoming aware of accidents and incidents to GA [general aviation] aircraft being flown by pilots holding professional licences and with several thousand hours flying experience .... Minding an Airbus while it trundles across the Atlantic does not qualify any pilot to safely operate a Pitts S-1C or a Rutan Long-Ez, to select two types featured in current AAIB [Air Accident Investigation Branch] accident reports' (Stewart-Smith 2005, p. 37).

In a 2003 *Log* article, Ron Ball gave his opinion on the pros and cons of highly automated aircraft (like the Airbus A320 family): 'Are [aircraft] not now incredibly reliable, comfortable and efficient machines that are a joy to operate on a good day? The answer is an unequivocal yes, but the flip side of that equation ... is that when they go wrong things can get very complicated very quickly and it isn't always straightforward to work out what's going on. So, for instance when the stick shaker goes off at Mach 0.85 at FL360 you can be pretty confident that you are not going to stall. You are less confident about what it's going to do next'. This kind of system opacity can give rise to latent errors. Under certain conditions latent errors may manifest as active errors (incidents or catastrophic accidents).

## Women's experience

As well as evolving technically, commercial aviation has evolved socially. One of the most notable intra-industry social changes has been the emergence of the female airline pilot. Women have made a significant contribution to both commercial and military aviation (a contribution that is often either overlooked or undervalued). During the Second World War, for example, female members of the UK Air Transport Auxiliary (ATA) ferried combat aircraft from their place of manufacture to front-line units. Women provided the same service in the United States through such entities as the Women's Auxiliary Ferry Squadron and Women's Flying Training Detachment. Gant (2003) says of women's contribution:

> As World War II progressed, women were able to break into many aspects of the aviation world. They served as ferry and test pilots, mechanics, flight controllers, instructors, and aircraft production line workers. At the beginning of 1943, 31.3 percent of the aviation work force were women. World War II was very beneficial to the movement of women into aviation fields. The history of aviation during these years is immense.

The end of the War, however, saw many women relegated to the domestic environment. Female workers were banished from the shop floor, commercial farm, railway, road haulage company and flight-deck. As Walker (2003, pp. 94–95) explains: 'World War Two provided a rare opportunity for British, American and Canadian women to prove their aviation abilities. British women pilots such as Pauline Gower, who commanded the women's section of the ATA ... fully expected to be able to continue their work as commercial pilots when the war ended. However, they were disappointed to discover that they were expected to return to the kitchen and became "The Forgotten Pilots" as detailed in the book of the same name by Lettice Curtis'.[46]

This pushing aside of women was one aspect of the gender dimension of Britain's post-war settlement. Hall and Jacques (1989, p. 26) note: 'Women had filled the labour market during the war, struggling for adequate nursery provision and against union agreements which ensured their employment was temporary. The patriarchal settlement ... was rehabilitated again after the second world war. Women were expelled from the labour market immediately after the war ... '.[47]

With respect to military flying, women had to wait some four decades to make a further contribution. In the UK, it wasn't until 1991 that the Royal Air Force commissioned its first operational female pilot (who commanded a C-130 Hercules transport aircraft based at RAF Lyneham in Wiltshire). In the New Millennium most RAF types, including the Jaguar, Harrier and Tornado (front-line combat aircraft), are flown by women pilots. In the USA, the Secretary of Defence removed the restriction on female combat pilots in 1994.

By 2005 the Pakistan Air Force was training women for combat roles. Given the conservative character of Pakistani society this was a significant development: 'Until recently, most women in this conservative Muslim society would more likely have imagined marrying a dashing fighter pilot than being encouraged to become one' (Abbas 2005).

While progress has been made, there is a feeling amongst female combatants that there is still much to be done. As a CNN correspondent remarked in 1999: 'Most ships in the [US] Navy now have women on board .... Many ... say integration will be complete only when people stop paying attention to them, when their numbers approach half those on a ship. That could take another 10 or 15 years' (Mintier 1999). Following the collapse of Communism in the early 1990s, military budgets were cut. While the 'peace dividend' provided for reductions in taxation and/or the re-allocation of resources, it had the effect of reducing the numbers of young people

46 The experience of wartime women pilots mirrored the experience of women in other occupations (like engineering and farmworking). Demobbed soldiers, sailors and airmen wanted their jobs back. Women were unceremoniously removed from their wartime occupations (as they had been at the end of the Great War).

47 It was not until July 2005 that the UK recognised in monumental form women's contribution to the defeat of Nazi Germany and Fascist Japan and Italy, when the Queen unveiled a sculpture memorialising women's wartime roles.

being trained by the military as pilots. This meant there were fewer time-served military pilots – both male and female – available to the airlines.

In the UK there is some disagreement as to exactly when women began to work as pilots for the majors. Walker (2003, p. 95) says: 'By the 1970s a handful of women were working as pilots for commercial airlines, most notably Yvonne Sintes, who joined Dan-Air in 1969, becoming the UK's first female pure-jet airline Captain flying a BAC 1-11 out of Gatwick. However, the credit for captaining Britain's first all-female crew goes to Captain Caroline Frost, who flew for British Air Ferries (BAF). The date was October 31, 1977, on BAF's inaugural flight from Southend to Düsseldorf'.

Barry (2005, p. 4) has a different view: '[W]hat of modern jet flying? In 1973 Emily Howell Warner became the first woman to be hired as an air transport pilot for a modern commercial airline and the first female Captain for a scheduled U.S. carrier. The first all-female crews were flying in 1978 (DHC-6) with FO Lynn Rippelmeyer and Captain Emily Jones leading an all-female flight deck and cabin crew'. The following account from Wood (2005, p. 38) sharply contradicts that of Walker and Barry: 'Dan Air ... started to recruit and train women from about 1961/2. I believe many women had very successful careers with Dan Air .... I think Sarah [Barry] does these women a disservice in ignoring them'.

What is not in dispute, however, is the fact that there are relatively few female commercial pilots. In the UK, for example, three percent of ATPL holders are women.[48] The figure is about the same in the USA for ATP holders. Barry (2005, p. 7) attributes this small percentage to a number of factors. First, the lack of positive role models for aspiring female pilots: 'Women often don't get exposure to flying and don't get to meet many pilots. When they do, those pilots tend to be male, white and middle aged and can be seen as being a little arrogant, macho and with strong views, which further enhances the perception that the job of airline pilot is only available if you have these characteristics'.[49] Secondly, there is still sexism in the industry. Thirdly, given that one tends to appoint in one's own image, the fact that most pilots are men produces a self-perpetuating hegemony of male pilots. Barry (2005, p. 7) observes: 'Until women start appearing in higher positions it is unlikely that management successors will be female'. How an organisation behaves is largely a function of its organisational culture (culture in this sense being 'the way we do things around here'). Organisational culture is shaped by the views and actions of those at the top of the management pyramid. It follows that if most (if not all) of those are men, women may experience difficulties in climbing the corporate ladder. Most airlines are still run by men.[50] The problem of the under-representation

48 In July 2005, women constituted 5.2% of BA's pilot group.

49 Interestingly, in her article on women in the Pakistan Air Force (PAF), Abbas (2005) found one recruit who said she had been inspired by her uncle, who had served in the PAF.

50 While there have been some notable exceptions (British Airways' low-cost offshoot Go, for example, was run by Barbara Cassani (until the airline was taken over by easyJet)) this is still very much the norm.

of women is not confined to aviation, however. It pertains in both UK political and corporate life. *The Guardian* (2005a) notes of UK industry:

> [J]ust one in 10 FTSE 100 directors' jobs are occupied by women. It is a truly shameful statistic .... A recent Deloitte survey showed that at FTSE 350 companies only 3% of executives and 8% of non-executives are women .... We have had the Kingsmill review, the Tyson report and Derek Higgs' urgings. We've had Patricia Hewitt [a Government Minister] pledging this, that and the other and now await the results of Lady Prosser's women and work commission set up last summer. Twelve months ago, in this column, we concluded *there was no explanation for the lack of senior women other than institutionalised corporate sexism at the most senior levels.* Nothing has changed to suggest we were wrong. (My emphasis)

Other factors also have a bearing on the number of female pilots. One consideration, for example, is the cost and availability of child-care facilities. Stephenson (2005, p. 37) observes: '[In Britain] it is easier to be a professional woman pilot if single or married with no children. To be married with children requires a good roster and a high degree of domestic organisation. Many women finally decide to place their home life and happiness above their airline career, even though it cost them an incredible struggle to achieve that career in the first place. Which is why there will always be fewer women pilots than men in present times'.

In 2005, a pilot at British Airways who had been denied fifty percent part-time working to allow her to look after her daughter decided to seek redress at an Employment Tribunal. The pilot's trade union stated: 'As the pilot workforce becomes more diverse, with more women entering the profession, a new range of employment practices is required including part-time working. We are stunned that British Airways hasn't woken up to this fact' (BALPA 2005b). The pilot herself[51] said: 'Being a professional airline pilot and a mother should in no way be incompatible ...' (Starmer cited in BALPA 2005b). In response, BA insisted that its decision not to allow fifty percent part-time working had been made on safety grounds *alone*. The airline argued that to capitalise on their training, low-hours pilots needed to build experience as quickly as possible. A BA spokesperson explained:

> Jessica Starmer currently works part time on a 75% contract .... British Airways has consistently said that the decision not to allow her to cut her working hours in half, to an average of just eight days a month, until she completes the required amount of flying hours was based on safety not sex discrimination .... British Airways believes that its pilots should have at least 2,000 flying hours experience – approximately three years of full time flying – before it is acceptable for them to work at 50% levels, which equates to only eight days a month .... This safety threshold is applied equally, whether the pilot is male or female. (Cited in British Broadcasting Corporation News 2005)

51 Jessica Starmer, 26, had obtained a first class honours degree from Oxford. After graduating she had secured a BA-sponsored traineeship. She was married to a BA pilot.

Having lost the initial Tribunal hearing, BA appealed. When BA lost the appeal the company said it would seek leave to approach the Court of Appeal, stating: 'This case is about safety *and British Airways' limited resources in the difficult business conditions that followed the terrorist attacks of September 11* .... The case is not about gender or part-time working' (my emphasis) (British Airways cited in Davies 2005). In mentioning economic considerations as a factor in their decision, BA would appear to be contradicting earlier statements to the effect that the decision was made on safety grounds alone (see BBC News (2005), above). BA seems to be saying in this later statement that the decision was made on both safety *and* economic criteria.

Hudson and Williams (1989, pp. 88–89) note: 'There are no simple outcomes from the dual roles that women occupy in domestic and paid labour; instead there are tensions between the two roles .... [T]he persistence of male chauvinism should not be underestimated'. Hall and Jacques (1989, p. 26) make this comment on the division of domestic labour between men and women in the immediate post-War period: '[W]omen's re-entry into the paid labour market [after 1945] was not matched by a transformation in men's relation to children, women or domesticity'. New Elizabethan Britain[52] was a determinedly patriarchal and conservative society. It is worth asking to what degree the attitudes of the 1950s and 1960s have changed? Writing in 1989, Hudson and Williams noted: 'The United Kingdom is a patriarchal society and this constitutes one of the major cleavages in British society. While there are a variety of definitions of patriarchy[53] these all centre on the notions of male domination and female subordination' (Hudson and Williams 1989, p. 88).

The advent of the two-crew cockpit has rendered the flight deck a less likely locale for sex discrimination, because there is no longer any possibility of a 'two-against-one' situation. Barry (2005, p. 4) explains: 'In the 2-crew environment ... discrimination is rare – probably because of the introduction and practice of CRM. If you do not like the person sat next to you, no one is going to be sympathetic to your view, so you must try to get on with them for the sake of a pleasant day at work'. A female First Officer for a holiday charter airline notes: '[Y]ou need to be able to find some common ground quickly for what could be a 10- or 11-hour flight. I can be quite shy, but on a one-to-one basis, I can usually find something to talk about' (McLaurin 2005).[54]

Both McLaurin and Barry believe that women bring different skills to the job. McLaurin notes: 'I think women bring a different approach to communication in the job. Discussions can be less confrontational and less ego-ridden. I don't think

---

52 That is, the Britain of Queen Elizabeth II.

53 Patriarchy is '... a system of social relationships through which men exploit women' (Hudson and Williams 1989, p. 88).

54 The two-crew flight-deck might be described as a mixed blessing. While concern has been expressed about pilots' increased workload (resulting from the elimination of the flight engineer), it would appear that this rationalisation has facilitated the inception of female pilots.

women make better pilots than men, but I do think the sexes bring different skills'. Barry (2005, p. 4) says: 'Although science tells us that women have different skills from men, all of these skills are needed on the flight deck, so for example, women are better at multi-tasking but not so good at spatial problems. With a male and female on the flight deck, this must surely increase safety'.

As to the public's reaction to the notion of a female airline pilot, McLaurin says: 'When I tell people what I do for a living, the reaction varied from "Gosh, you must be brave and clever!" to "What, you're an air hostess?" The majority of people still expect pilots to be middle-aged, white males and ex-RAF. Often my occupation stops conversation completely ... '. Barry (2005, p. 4) notes: 'Female pilots are often on the receiving end of patronising comments from the passengers – comments such as "well done for getting us here" or "when will you be able to fly the plane?" or "you weren't flying were you?" are common'. Barry (2005, p. 4) concludes: '[F]emale pilots need to be diplomatic and resilient ... '.

Recognising that women are often deterred from entering the industry some of the more enlightened employers have launched major initiatives to attract women. In the manufacturing sector, for example, European Aeronautic, Defence and Space (EADS) runs a Women's Day at the Paris Air Show. An EADS personnel manager has noted: '[Women] know that aerospace is a pretty male industry and they are sceptical about how they can work in it or integrate with it' (Gonce cited in *Flight International* 2005, p. 65). EADS says there are benefits in engineering mixed-gender project teams. Women bring different skill sets and outlooks to work, it says.

## Media constructions

Commercial aviation attracts media attention for a number of reasons. Journalists are seduced by its glamour, technical achievements (Concorde, for example), economic and cultural impacts, and by the fact that, for some, flying constitutes a 'dread risk'. According to Bor (2003) a significant minority of passengers are nervous about flying. For some the prospect of trusting flight and cabin crew, air traffic controllers, engineers and other aviation professionals with their destiny is a source of fear and trepidation (so much so that normally restrained individuals take refuge in the bottle or pill-box).

Depending upon the circumstances, flight and cabin crew are portrayed as either heroes or dolts, Madonnas or whores. While there is disagreement over the amount of direct influence exerted by the media, it is accepted that the media help shape society's agenda of concerns. For this reason, the way that journalists portray the industry should be of concern to all those directly employed by, and indirectly dependent upon, the world's airlines.[55] The following six newspaper accounts serve to illustrate the extremes of media representation.

---

55 It is estimated that commercial aviation sustains some 500,000 British jobs (Bennett 2002b).

*The Madonna: case I*

In May 2000, a British Regional Airways BAe ATP twin-turboprop aircraft *en route* from Manchester to Belfast made a forced landing at Liverpool John Lennon Airport with a landing gear fault. The left main gear collapsed on touch-down. The story, which took up the whole of the front page of the May 17th edition of the *Manchester Evening News* (a tabloid-format[56] regional newspaper), carried the banner headline: 'Terror-flight passengers hail Captain Fantastic'.

The report reproduced several statements from passengers. One passenger said: 'The captain was fantastic and the stewardesses were great, even though they must have been panicking inside. I'd like to thank him for what he did. He was brilliant'. Another said: 'It was terrifying. About 10 minutes after take-off the pilot announced there was a technical problem. We just waited – and waited. It was the longest three hours of my life up there. But the Captain and the crew were superb. We just knew we would all be all right'.

The Chief Executive of British Regional Airways commented: 'All pilots have extensive simulator training but of course there is a big difference between that and the real thing. We have not had a chance to say very much to him ... but it is clear from feedback we have had from passengers that he has acted with great calm'. The story took up the whole of the newspaper's front page and all of page 3. Page 3 carried the headline: 'Cheers for terror-flight pilot'.

*The Madonna: case II*

In November 1997, a Virgin Atlantic Airbus A340-300 made an emergency landing at Heathrow with its left landing gear only partly extended. The gear collapsed on touch-down. The *Daily Mail* (a tabloid-format national newspaper) carried the story on its front page under the banner headline: 'Inches From Catastrophe. After an extraordinary series of mid-air moves, pilot dramatically lands crippled jet at Heathrow'. The story took the entire front page and the whole of page 5. The story quoted one passenger as saying: 'The pilot was wonderful'. One of Heathrow's Chaplains was interviewed: '[The passengers] are all very relieved and full of praise for the captain. Everyone is singing his praises. They say he did a wonderful job'. The *Mail* journalist noted: 'Plane enthusiasts listening to the conversation between the controllers and the pilot described him as "incredibly calm" ...' (Williams 1997). He also observed: 'Captain Barnby, whose father and grandfather were also pilots, has flown some 14,000 hours in various aircraft, including gliders, Boeing 747s and Spitfires'.

Both this article and the item in the *Manchester Evening News* constructed the pilots as quiet, modest professionals. Having quoted a grateful passenger, the *News*

56 In the UK, tabloid-format newspapers (like *The Sun*, *Daily Express*, *Daily Star*, and *Daily Mail*) are taken to have certain characteristics, such as the use of blunt, often hyperbolic, attention-grabbing headlines and (necessarily) abbreviated analysis.

article reproduced this statement from the pilot: 'The passengers deserve the credit. They didn't panic and made my task easier'. The *Mail* article noted: 'Captain Barnby played down his achievement'.

*The Madonna: case III*

In August 2001 a British Army Aircraft Accident Board of Inquiry published its final report on the crash of a Lynx helicopter in Leicestershire in May 1999. The crash pilot survived with serious injuries, but three others died. The *Leicester Mercury* (a tabloid-format regional newspaper) carried the story of the Inquiry report on its front page under the banner headline: 'Air Crash Hero Saved A Village. Bravery of helicopter pilot in accident that claimed three lives'. The report quoted the local MP as saying: 'The pilot was clearly a hero'. A local resident was quoted: 'If the pilot kept the helicopter away from homes and buildings, then he should be congratulated'. The *Mercury* story commenced with this eulogy: 'The bravery of a pilot who steered his burning helicopter away from a Leicestershire village emerged today .... Villagers at Tilton-on-the-Hill praised the pilot who flew the Lynx to clear ground, missing homes and farm buildings, seconds before it crashed' (Newton and Ridley 2001).[57] The story occupied the entire front page.

A common denominator in all three accidents was the skill demonstrated by the pilots. To invoke Tom Wolfe, it is clear that each pilot had sufficient quantities of *the right stuff* to be able to overcome the Fates. In his book *Redefining Airmanship* Kern (1997, pp. 21–22) says that pilots with *the right stuff* '... understand the capabilities and limitations of themselves; their team; their aircraft; the physical, regulatory, and organisational environment; and the multiple risks associated with the flight. An expert flyer builds on a bedrock of flight discipline, skills, and proficiency'.

*The whore: case I*

In September 1999, a Britannia Airways Boeing 757 carrying 245 passengers and crew broke into three sections after skidding off the runway at Gerona. The aircraft had landed in a storm. Other crews had decided to divert to alternates. There were no fatalities, although some passengers received minor injuries. The British national tabloid *The Sun* ran the story on five pages, including the front page. The front page carried a photograph of the crashed aircraft under the banner headline: 'We told him not to land'. The six contributing journalists noted: 'Some passengers ... blamed the skipper. But most PRAISED him and the crew for averting a disaster'. One passenger was quoted as saying: 'The crew were brilliant and the captain, who was covered in blood from a deep head wound, went round asking how we all were'.

57 Of course, the Army pilot's *primary* objective in flying to clear ground may have been to attempt an emergency landing (thereby increasing his chances of saving his comrades and the stricken Lynx). Regardless of this possible motive, the *Mercury* constructed his actions as a heroic and public-spirited attempt to save the lives of rural civilians.

Another said: 'The pilot tried to land twice. I thought he should have turned back but he didn't. I think he thought he was Biggles. The plane kept plopping up and down and finally we dropped like a stone on to the runway'.

The journalists admixed three other aviation-related stories with the main coverage. The first concerned an aircraft that had skidded off a runway at one of Rome's airports; the second reported on Spain's stormy weather; the third reported on a near-miss between two aircraft near Luton Airport, just north of London (this incident had actually occurred on 14 August, 1998).[58]

What is most interesting about *The Sun*'s reporting is the fact that although the article reproduced several statements in which the flight crew were praised, the journalists used the headline: 'We told him not to land' (a reference to the Spanish air traffic controllers' version of events) on the front page (Parker *et al.* 1999).[59]

### *The whore: case II*

In June 2002 the UK media, including *The Times* newspaper (a national broadsheet), covered a report from the Confidential Human Factors Incident Reporting Programme (CHIRP) concerning some pilots' responses to air traffic control instructions. *The Times* carried a report on its front page under the headline: 'Budget airline pilots accused of putting safety at risk'. The report commenced with the following summation: 'Pilots working for some low-cost airlines are disobeying air traffic control instructions because they are under extreme pressure to meet tight schedules, according to a confidential safety report' (Webster 2002). The article went on: 'The report says that controllers are receiving "overly aggressive responses", with pilots repeatedly challenging information on visibility and on whether the aircraft in front

58 Journalists often group stories together to accentuate the 'spin' they put on the main item. For example, in the 5 August 2005 edition of *Metro* (a free newspaper circulated in Britain) page 11 carried three aviation-related stories, each of which concerned some aspect of flight safety. The main story dealt with an aircraft diversion due to smoke in the cabin (Stallwood 2005, p. 11). The 'second feature' dealt with the crash-on-landing of an Airbus A340 (where fire destroyed the aircraft but all 309 people on board escaped). The third feature dealt with a passenger's attempt to open an aircraft door while in flight. The first event occurred in the UK, the second in Canada and the third in the USA. The main story, titled: 'We thought we'd die on smoke-filled plane', carried a reference to the Airbus crash. It read: 'The incident came just hours after the Toronto crash, where an Air France plane skidded off the runway and burst into flames'. Given the number of flights made each day around the world it is unsurprising that two aircraft should encounter a smoke/fire problem on the same day. The linking of these stories, however, served to create an impression of crisis. (Of course we cannot be certain what, exactly, *Metro* was trying to achieve by clustering the stories.)

59 The front page carried a second, unrelated story along its foot under the headline: 'Viagra Doc Held. He uses lover in sex experiments'. British tabloid newspapers' agenda of concerns is relatively limited. The focus tends to be on stories concerned with gruesome death, sex, corruption or sporting achievement (generally football/soccer). Shock and titillation are used to boost sales.

has safely cleared the runway'. The journalist said he understood that the report referred principally to low-cost carriers. Three low-cost carriers (Go, easyJet and Ryanair) were reported as saying that safety was their number one priority.

### *The whore: case III*

In June 2002 the *Daily Mirror* (a British nationally-distributed tabloid newspaper) carried a report about a cabin crewmember who mistook dry ice[60] fumes for a smouldering fire. The *Mirror* covered the story on page 24 of its 26th June edition under the headline: 'Air Girl's Fire Alert Over Dry Ice Fumes'. The story commenced: 'An air stewardess mistook a white plume from dry ice for smoke and started a fire alert that led to an emergency diversion'. The crewmember discharged a fire extinguisher onto what she thought was the seat of the 'fire'. The report concluded: 'A replacement plane was sent to pick up the ... passengers, costing easyJet thousands' (White 2002).

Regarding the above, three points can be made.

### *Aviation stories make good copy*

Because aviation fires the public imagination (in both positive and negative directions), journalists know that a good airplane story told with gusto will attract readers – which is one of the major objectives of *all* journalists. Bonner (2003, pp. 14–15) notes: '[N]ews organisations are often accused of ignoring the good to concentrate on the bad .... Well, yes. Who, after all, is going to rush out and buy a newspaper ... to learn that everything worked the way it should that day? Reporters ... focus on things that break down, crash or fail through accident or design – because that is what people are interested in. It sells newspapers and swells audiences'.

### *Hyperbole is a norm*

Because newspapers – especially the tabloids – focus on the dramatic rather than the mundane, flight and cabin crew involved in incidents are portrayed as either heroes or villains. Flights that pass without incident due to crewmembers' professionalism are, in media terms, worthless. Aviation reporting consequently see-saws between two unedifying extremes: crewmembers are either canonised or demonised. They can also be 'idiotised', as with the *Mirror*'s construction of the easyJet cabin crewmember, above. The *reality* is that cabin crewmembers are trained to put safety before all other considerations. Clearly this is what the crewmember did. She took what she believed to be a reasonable precaution. The *Mirror*'s reference to the cost of the diversion and failure to mention the 'cautionary principle' evidences the shallowness of tabloid reporting.

---

60 Dry ice is used in galleys to keep food cool.

### *Mixed messages – an emergent property of popular reporting*

The media's often hyperbolic reporting on commercial aviation is dangerous for two reasons. First, because it obscures the fact that aviation is one of the safest forms of public transport. Secondly, because it *may* serve to undermine the confidence of happy fliers, and destroy the diminished confidence of nervous fliers. As mentioned earlier, there is disagreement as to the influence exerted by the media on public perceptions. Nevertheless, given the possible negative outcomes of augmenting fliers' unease, is it not advisable to avoid sensationalism when reporting on commercial aviation? Commercial aviation creates wealth, facilitates cultural exchange (and therefore promotes understanding) and encourages innovation. For many employees, aviation is more a vocation than a job.

Many years ago, the Chairman of BOAC penned the following on the subject of the social impact of the Comet airliner (the world's first jet-powered passenger aircraft): 'If the new jetliner is good to look upon so, in my opinion, is the purpose for which it has been set in the firmament. Its great speed has shrunk the world to half its former size and foreshortened the frontiers in many countries across the globe. It has created a new vogue for international travel and this, I hope, will help to create a new pattern of international living and understanding. I believe the Comet to be a symbol for peace, progress and prosperity' (Thomas cited in Hensser 1953, p. 5). Those responsible for the 1944 Chicago Convention expressed the same hope. The Convention stated: '[T]he future development of international civil aviation can greatly help to create and preserve friendship and understanding among the nations and peoples of the world ...' (cited in International Civil Aviation Organisation 2003, p. 13).

If commercial aviation is a social good (see in addition to the above the argument made by Bennett 2002b),[61] then it deserves a fair hearing. While it would be wrong (and socially dysfunctional) to view the industry through rose-tinted spectacles, one thing is clear: the media's schizophrenic reporting does the commercial aviation industry (and the many dedicated people who work in it) a disservice.

## Summary[62]

The aviation industry has two defining characteristics. First, it exists in a state of permanent and aggressive technological, operational and organisational revolution. Secondly, despite the industry's revolutionary character, pilots' professional ethic

61 Economic and cultural benefits come at a cost, however. Aviation causes environmental disamenity. Learmount (2005, p. 13) notes: 'The UK's aviation industry admits growth in air travel "has created significant detrimental environmental impacts ..."'.

62 It is important to remember that the points made here are made on the basis of *secondary* data (that is, information garnered from the work of others). As to the status or weight of this type of data, Macdonald and Tipton's (1993, p. 187) comments are worthy of reprise: 'The tradition of documentary studies in sociology started by these eminent scholars

– that determination to do the best job one possibly can – remains unchanged. It is one of the industry's few constants, transcending both time and national culture.[63]

Over the years, the world's airlines have had to accommodate many changes. It has been (and still is) a rollercoaster ride. Some of the more significant trends and events that have impacted the industry are as follows.

*The decline of deference*

This long-term social trend has meant that authority figures, including pilots and cabin crew, are no longer fêted. Indeed, it is now the fashion to debunk or vilify 'errant' pilots and cabin crew (see above). The debunking may take place before the full facts are known (often for the most ignoble of reasons, like selling as many newspapers as possible, to make as much money as possible).

The industry itself has also played a part in the erosion of deference. The introduction of crew resource management, for example, has served to flatten (but not eliminate) the flight deck's authority gradient. To improve safety margins much of the imperiousness of the rank of Captain has been engineered away. First Officers (and flight attendants) are encouraged to contribute to the management of the flight and to seek clarification from the Captain if, on the basis of a piece of information known only to themselves, they believe a decision to be ill-advised.

*The sunset of imperialism*

Countries' foreign adventures spawned numerous flag carriers, such as Air France, KLM, BOAC and Pan American (although Pan Am was more chosen instrument than bespoke flag carrier). The retreat from empire in the 1950s and 1960s marked a sea change in flag carriers' fortunes, which eventually impacted their employees.

*The imposition of order*

For the first two decades of its existence, the commercial aviation industry bore more than a passing resemblance to the Gold Rush of America's mid-West. Rather like an atomic explosion, the Wright Brothers' first flight unleashed an enormous amount of energy. New designs were brought to market and new services started with dizzying regularity. The resulting chaos prompted governments to intervene. Keynesianism, the political zeitgeist of the 1930s, migrated to commercial aviation.

---

[Marx, Durkheim and Weber] has not declined over the years .... Documentary research remains an important research tool in its own right ... '.

63 Helmreich and Merritt (1998, p. 70) note: '[A]ll pilots ... share a professional culture that ... transcends the influence of national culture in many important ways .... Flying brings with it certain behavioural necessities and traditions that create a culture unique to pilots around the world'. Kern (1997, p. 1) observes: '[F]lyers have a separate professional identity beyond their official rank or aeronautical rating ...'.

In the US the CAB was set up to introduce a measure of order to the industry. Routes were allocated and flying hours regulated. Eventually, supra-national bodies like ICAO and IATA were established (although states could 'contract out' of ICAO's Standards and Recommended Practices (SARPS)).

Today, commercial flying is a deadly serious endeavour. The barnstorming days are long gone. Aviation is a highly regulated activity. It is mathematised and mechanistic. Almost everyone in the industry – from cleaners to pilots to cabin crew – understands, conforms to and reproduces this culture. The following statement from a UK-based charter captain shows the degree to which flying has been routinised, and the role of the pilot transformed: 'I ... fly between 600 and 900 hours a year. This may not sound much, but additional time is spent on pre-flight briefings in the crew room and setting the aircraft up before departure, which includes on-board computer aids to be used. In addition to being responsible for the passengers' safety and flying the aircraft, I have to deploy management skills and motivate, monitor and be responsible for the actions of my crew – both in the air and on the ground' (cited in Prospects.ac.uk 2005). Scott's (2005) evaluation of the modern cabin crewmember gives us a further insight into the culture of commercial aviation: 'Becoming an "air hostess" was once the height of glamour for young working-class girls ... they got to see the world .... But that was the 1960s and the lot of the trolley dolly has changed: political correctness put an end to age restrictions and the responsibilities go beyond asking passengers whether they want "chicken or fish". Cabin crew, as they are now known, must be schooled in all manner of safety regulations and must receive first aid training. The recruitment process is long and drawn out ... '.

*The organisation of labour*

After the chaos of commercial aviation's early Gold Rush, its workers began to organise, both nationally and internationally. Terms and conditions of employment improved. Eventually the pendulum swung so far in favour of organised labour that pilots could demand pilot-engineers on the flight deck.[64] In the US the pendulum remained in favour of pilots even after deregulation. Following industry upsets such as the terrorist attacks of 11 September 2001 some struggling airlines looked to pilots (and flight attendants) for major concessions on terms and conditions.[65]

64 The head of the United Airlines pilot union once remarked: 'We do not want to kill the goose, but want to squeeze every last golden egg out of it' (Dubinsky cited in Field 2005b, p. 30). In the US the apogee of union power arrived in 2001 when a United Airlines senior Captain could earn between $198 and $265 an hour (Field 2005b, p. 30).

65 Writing in August 2005, Field observed: 'Hacking away at payroll and slashing benefits have been the first tools at hand for most airline managements .... Collectively [airline labour has] given concessions worth $6 billion .... At US Airways, some pilots have seen their salaries fall from $150,000 in 2002 to $70,000 a year now. One United flight attendant has said publicly that between 2000 and the second quarter of this year, her annual salary fell from $52,000 to $39,000' (Field 2005b, p. 30).

*The imposition of disorder*

The *status quo* established in the 1930s lasted until the late 1970s, when the US deregulated its industry. Deregulation saw the lowering of barriers to market entry. There were many start-ups ... and many failures. The launch-flop-rationalisation cycle seems to have become a permanent feature of the industry.[66] Today, even countries with a long history of protecting their chosen instruments are surrendering to the forces of market liberalisation. Protectionist India, for example, '... has acted on liberalisation pledges and significantly eased restrictions in air services accords with key trading partners ... ' says Ionides (2005, p. 22). India has also deregulated its domestic market. As of July 2005 the country had three low-cost carriers (Kingfisher, SpiceJet and Air Deccan) with more LCCs planning to launch. The buoyant Indian market has attracted the attention of British Airways, which in November 2005 launched a new service to Bangalore. BA's Chief Executive Officer said he planned to launch two more new services – to Chennai and Hyderabad. Asked why he was concentrating on India instead of China (the world's largest and fastest-growing economy), BA's CEO said: 'India is much more developed in terms of travel culture and tourism. BA has been flying here for the past 72 years, so we know them and they know us' (Walsh cited in Kane 2005). According to Kane (2005): 'BA sees India as its second most important market, after North America'. Clearly, established airlines like BA with well-rehearsed operations will seek to exploit promising new opportunities.

The liberalisation-and-deregulation bandwagon has penetrated even the Middle East. However, in a region where change happens at barely a snail's pace, liberalisation is very much a long-term project. As Endres (2006, p. 25) explains: 'It could take five years for some of the current privatisations to be completed successfully .... Liberalisation ... will take much longer to achieve. IATA director general Giovanni Bisignani succinctly summarised the current situation by saying "Change is critical, but it is slow in coming"'.

In Africa, the 1999 Yamoussoukro Decision '... set the framework for the intra-African liberalisation of the airline industry' (Chingosho 2005, p. 114). Recent years have seen the establishment of several LCCs (that are able to offer travel opportunities to a larger proportion of the population, thereby promoting economic development) and the privatisation of some national airlines. Chingosho (2005, p. 114) asserts: 'Government focus on good political and corporate governance and liberalisation programmes auger well for the future'.

In Europe, deregulation has become established as a central tenet of government. In Germany, says Wegrich (2005, p. 4): '"Entbürokratisierung" (de-bureaucratisation)

---

66 Interestingly the country that spawned the deregulation movement – the United States of America – has begun to raise barriers to market entry with the intention, says one market analyst, of 'weeding out the kooks [no-hopers]' (Jenkins cited in Field 2005a, p. 13). It is possible that as other countries' markets begin to oscillate, they too will seek to restore equilibrium by 'raising the bar'.

is once again high on the agenda' (although 'powerful anti-deregulation coalitions' have emerged in response). During the 2005 British General Election, according to Hutter and Power (2005, p. 3), '... the regulatory burdens upon industry became a key area of debate between the main political parties'. Prior to the election: '[T]he Chancellor of the Exchequer outlined his plans for cutting red-tape and increasing the efficiency of regulatory enforcement by publishing and endorsing a report by Sir Philip Hampton, *Reducing administrative burdens: effective inspection and enforcement* [published in March 2005], and by proposing a new Better Regulation Executive to implement Hampton's proposals' (Hutter and Power 2005, p. 3).[67]

### *Long-term growth*

There is a paradox at the heart of commercial aviation: regardless of circumstance, regardless of the chaos around it (and occasionally within it) the industry continues to grow. While regional conflicts, terrorism, energy crises, environmental disasters, epidemics and other perturbations do hurt the industry, the effects are invariably short-lived. The industry takes the pain, adapts and moves on. This says something about the calibre of the people who work in it, and the marketability of its products. One should not underestimate the impact of this volatility on employees' psyche, however. Pilots' love affair with flying makes redundancy hard to bear.[68]

### *The technologisation of flying*

While most early pilots yearned for better instrumentation and a more comfortable working environment, there is a feeling that, with today's highly automated glass cockpits, pilots are being disempowered and deskilled to the point where they may become a liability (due in part to a degraded situation awareness (SA)). It remains to be seen whether the shift in the 'balance of power' between the pilot-craftsman and technologist will have any long-term implications for the profession.

The deskilling argument is not as coherent as it might seem, however. It *could* be argued that pilots are merely being required to deploy a different – but no less important – set of skills (like real-time system monitoring and management, with occasional problem-solving under conditions of physical and psychological stress). According to Wood (2004, p. vi), however, advanced automation may degrade pilots' manual flying skills to the point where they are no longer able to intervene successfully: 'There is an increasing amount of automation in aircraft and greater reliance on this automation by the crew. Consequently, there is a risk that flight

67 Following the widely publicised deaths of 21 cocklers in Morecambe Bay in 2004, the British Government created the Gangmasters Licensing Authority – a new administrative function. In light of this it could be argued that the British Government's policy on deregulation is not as coherent as it could be. The lack of coherence in policymaking may result from politicians' sensitivity to media pressure.

68 It also makes retirement hard to bear.

crew no longer have the necessary skills to react appropriately to either failures in automation, programming errors or a loss of situational awareness'.

In his review of cockpit automation over the last two decades, Dekker (2004, pp. 183–184) judges automation to be something of a curate's egg. He notes: 'The introduction of automation over the past twenty years has subjected air transport to a sizeable natural experiment.[69] Its results form a pattern that is quite different from the original promises .... The benefits of automation are most noticeable when they are of least value: during low-tempo operations when the human operator already had little to do. This while the burdens of automation (its demands on human memory, cries for more input, difficult-to-manage interfaces, etc.) accrue during high-tempo, safety-critical operations, when the operator had always been busy with other tasks and where margins for error were smaller and time pressure higher. This pattern [has been] called "clumsy automation" ... '.

*The advancement of women*

Despite commercial aviation's historic maleness, women have succeeded in advancing their position and status within the industry. Unfortunately, however, the failure of a still-patriarchal society to cater adequately for women's needs means that female pilots who are also mothers have difficult career choices forced upon them. Women also have to come to terms with travellers' sexist attitudes (the term 'trolley dolly' is still to be heard) and the persistence of sexism amongst a small number of flight crew. It would seem, however, that technologisation may have helped women's advancement (the two-person flight deck promotes teamworking and makes a two-on-one situation impossible).

*Twenty-four-hour scrutiny*

The world's new satellite and Internet-driven mass media has a voracious appetite for news. Some sections of the media (the British tabloids, for example) are more concerned with sensation than fact-finding. This can harm the industry and its employees, including pilots. Pilots involved in incidents or accidents may be subjected to intense media scrutiny (as might their families). The mass media often seeks instant answers to complex questions. Speculation alarms passengers and damages the industry (but it can help sell newspapers).

69 The notion that areas of social and economic life can become a 'living laboratory' for the scientist and technologist has been put by others. See, for example, Ulrich Beck's analysis of the relative autonomy of scientific experimentation in his book *Risk Society* (1992). Beck argues that because of its global mobility, scientific experimentation is largely beyond the control of individual nation states. The result? A global laboratory in which each individual is a candidate for experimentation.

*The resilience of pilots' professional ethics*

Like any professional group, pilots seek to do the best job they possibly can under the prevailing circumstances. It matters not whether they are flying a new Boeing 747-400 between London and New York for a glamorous trans-Atlantic carrier, or a 20-year-old DeHavilland Twin Otter in Alaska for a mining outfit, the objective – to convey the load safely and efficiently from point A to point B – remains the same. To do this pilots draw on their training and accumulated skills and experience. They also use teamwork, drawing on the skills and experience of co-pilots, cabin crewmembers, air traffic controllers and ground-based engineers (this is the essence of crew resource management (CRM)).

Like test pilots, line pilots must also be able to dig deep when things go wrong. They must have sufficient of *the right stuff* to solve ill-defined problems under conditions of physical and psychological pressure. *Every* commercial pilot needs this facility, because computer programmers make mistakes when entering and checking code, microprocessors and transistors cook themselves with the heat they generate, insulated cable degrades over time, aircraft suffer mechanical failures, air traffic controllers err and storms create hazards for even the most advanced aircraft.[70] Pilots' professional ethics are a constant in an industry defined by its inconstancy. Commercial aviation changes. Pilots fundamentally don't.

70 In August 2005, for example, an Air France A340-300 crashed on landing at Lester B. Pearson International Airport during a heavy rainstorm. There was a suspicion that the aircraft had aquaplaned off the end of the runway. While all 309 persons on board escaped, the aircraft, which skidded in to a shallow ravine, was destroyed by fire (Reid 2005, p. 25). Investigators discouraged speculation as to the exact cause of the crash.

## Chapter 3

# Pilots' Socio-economic Origins and Motivations

The questionnaire probed pilots' social and economic background. For example, it asked respondents whether they had attended nursery school (that is, pre-school or kindergarten) and to describe their parents' occupations. This multi-level analysis produced a high-resolution socio-economic profile for each respondent. The profiles are aggregated in Appendix 3: Pilots' Socio-economic Origins and Motivations. In this study of socio-economic origin, class is defined in terms of occupation (a crude but, in this author's opinion, tractable definition).[1] Examples of working class occupations would be a school caretaker or farm labourer. Examples of middle class occupations would be a teacher or farmer. Examples of upper-middle class occupations would be a bank manager or hospital Registrar.

Applying this algorithm to the questionnaire data it is seen that almost three quarters of respondents were born into middle or upper-middle class households. Households-of-origin were relatively small in size, with only eighteen percent having more than three children. It has been noted that as societies develop and become more prosperous, family size decreases (as fewer children are lost to preventable diseases, for example).

There was relatively limited social mobility amongst respondents' parents. Middle or upper-middle class parents lived longer than working class parents. In the UK it has been noted that: 'There are wide variations in life expectancy by social class'. In 1999 the life expectancy for members of the 'professional social classes' was 82.8 years for females and 78.5 years for males. In the same year the life expectancy for the 'unskilled social classes' was 77.1 years for females and 71.1 years for males (South West Public Health Observatory 2005).

A significant minority of respondents had attended university. Those born into working class households were less likely to have attended university than those born into middle or upper-middle class households. Summerfield and Gill (2005, p. 36) note: 'Young people in manual social classes are under-represented in higher education in Great Britain and the gap in participation rates between those from manual and non-manual classes has increased over the last 40 or so years ... '. One

1 Harvey and Macdonald (1993, p.10) note: 'This approach ignores all the other aspects of social class and reduces class to *occupational role*. Such an approach derives from Durkheim's work (Durkheim, 1895, 1897, 1947)'. Of course, class is but one social cleavage. There are others. For example gender, age, race and religion.

respondent who had grown up in the Midlands in the 1950s explained that the option of attending university had never been available to him. His father had worked as a railway parcels delivery driver. His mother had been a housewife. There were few college bursaries (indeed, there were few colleges). His only chance of climbing the social pyramid had been to get in to the Royal Air Force. As he explained during his interview:

> I started off as a salesman [his first job had been with Hoover Ltd. as a Business Trainee] so I would probably have finished up running some business. Later in life I realised I wasn't as academically stupid as I thought I was at school, and that a whole range of jobs would have been open to me. I have always enjoyed the law and medicine and things like that .... Of course as a young kid from the Midlands [Nuneaton] in the 1950s those careers just were not available. My parents did not have the money to subsidise me .... It was very good fortune that I managed to get in to the RAF at the right time ... that was my big turning point ... I do not know whether you saw the film *An Officer and a Gentleman*? In it the drill sergeant asked Richard Gere: 'Why do you put up with this?' to which the main character replied: 'Because it is all I have got'. That really summed up, at least for the more honest ones, how it was for us in the RAF ... most of us were not well-qualified ... we used to say: 'We cannot give this up and go and be a brain surgeon or a concert pianist ... this is it ... we have got to make the best of this'. As it happens, we enjoyed it.

Pilots' career paths were far from linear (that is, school to university, to flying scholarship or military flying, to commercial flying). Many respondents had come to commercial flying from unrelated occupations. One had been a Trainee Quantity Surveyor, then a milkman, then a Production Operator for the Ford Motor Company. This respondent, a male, cited 'no interest in the job' as the reason for giving up on surveying and milk delivery. Another had been a Post-Doctoral Research Assistant (at the universities of Bristol and Nottingham), then a Staff Scientist (for a biotechnology company). Another had worked as an Assets Manager and Capital Markets Manager (for various companies including Chase, UBS and Citibank). This respondent cited boredom as the reason for leaving some of his finance jobs. Other career paths included:

- Branch Manager for a firm of electrical factors, then an Export Sales Manager for another electrics company.
- Bank Clerk at a high-street bank, then a Marketing Assistant for the same bank, then a long-haul cabin crewmember. Her first flying job was with LCC Debonair, which she left for Sabena (citing better pay).
- Laboratory Technician at the University of Cambridge.
- Administrator for the London Borough of Islington.
- Scientific Officer at a transportation research establishment. On leaving this job the respondent became a tug pilot at a gliding club, then a flying instructor.
- Estate Agent then a General Office Manager.
- Immigration Officer for the UK Home Office.

- Showroom Salesman for an electrical retailer then a Salesman for a marine outlet.
- Saleswoman for a computer software company, then an Outward-Bound Instructor.
- Apprentice Toolmaker (at British Aerospace), then an Apprentice Mechanic, then a labourer. This pilot cited 'lack of interest' as the reason for his leaving British Aerospace.
- Electrician, then a Mechanic-Electrician, then an Engineer (for Swissair Technics).
- Hydrographic Surveyor.
- Warehouseman (for Rand McNally).
- Flight Attendant (for Western Airlines).
- Flight Attendant (for Air Atlanta), then an Operations Officer for the same airline.
- Sailor (in the Swedish Merchant Fleet), then a truck driver.
- Office worker, then a shop worker (for the UK supermarket chain Asda).
- Engineering labourer.
- Deckhand, then a fishing boat Captain.
- Salesman (for a biscuit manufacturer).
- Bar Manager (at various five-star hotels in Australia). This respondent left the hospitality trade to work for a skydiving club.
- Telecommunications Technician (at Eircom, the Eire telecommunications company). The respondent joined Aer Lingus after leaving Eircom.
- Development Engineer for British Telecom (BT). The respondent became a commercial pilot because 'Commercial flying seemed more interesting than boring BT'.
- Aircraft Cleaner (at Air Atlantique). This respondent left school with eight GCSEs and two A-levels.
- Hotel Apprentice (kitchen and dining room). This respondent stated his reason for leaving as: 'I was confused'.
- Apprentice professional footballer (for Sheffield Wednesday).
- Outward-Bound Instructor, then a Sports Event Manager.
- Stewardess (for Pan Am then Lufthansa). This Berliner flew with Pan Am until the airline folded.
- Engineering Officer in the British Merchant Marine then a Metropolitan Police Officer.

One female interviewee explained that she had been made aware of flying as a career by some of her trainee paragliders:

> [I became a commercial pilot] by chance. I was a paragliding instructor. I enjoyed it because it got me out of doors ... and I like communicating and being sociable with people ... but I was always conscious that it would not keep me going for forty or so years. I knew the career was only temporary ... but I had no idea of where I should go .... Some of my paragliding trainees were pilots. They offered to take me on the jump seat. So I

> got a couple of jump seat rides. After my third jump seat ride back into Manchester the pilots asked me whether I enjoyed it and I said: 'Well, I could do that'. I wasn't being big-headed but I knew that I could do it. After that I decided I wanted to be a pilot. It was so blatantly obvious.

This interviewee, who had worked for a time in computer sales, stated: 'I never had a great plan at school. My only careers advice was: "You are too good to be a hairdresser"'. Despite the paucity of careers advice – and possibly of encouragement, too – the interviewee went on to obtain an Ordinary National Diploma in Business Studies and Higher National Diploma in Computer Studies at Trent Polytechnic. It is worth pondering whether the interviewee's school was guilty of gender stereotyping. After flying for a UK-registered airline for some four years, she left to take a management job in the National Health Service (NHS). Using her knowledge of CRM she did pioneering work on team building with nurses and doctors. She also obtained a Masters degree in Risk, Crisis and Disaster Management from a British university.

### *Flying: job or vocation?*

For the ex-paragliding instructor, flying was never anything more than a job – a means to an end:

> [Flying was] a job because I flew to maintain the lifestyle that my husband and I wanted. It was good. I enjoyed it. But I did it so we could get on with doing the things we wanted to do in our lives .... It was an exciting and enjoyable means to an end.

Asked to define vocation she said: 'A vocation is something you are committed to and believe in ... something that is within your soul or psyche, rather than just being a job'. Another interviewee stated: 'A vocation is something you do ... for the love of the job, be it well-paid, unpaid or whatever'. She continued: 'I think [commercial flying] *is* a vocation because there are so many things that are bad about it that if you did not enjoy it you would not do it. The rewards are there as well, but people on the street do not realise how difficult it is physically, relationship-wise and all the rest of it. I think it has to be vocational'. A retired male pilot said: '[Flying] tends to be a job when you are in it and a vocation when you want to get in to it, or when you are out of work. I do not think it should be a job. One of the reasons I retired at sixty was because commercial flying was becoming more and more of a repetitive job, and I think it needs a degree of enthusiasm because you are there dealing with people in the crew and if you really are fed up you should not be in the crew. So it can become a job ... but it should be a vocation'. The importance of maintaining one's enthusiasm for flying was emphasised by another pilot:

> [Flying is a vocation] because you cannot do it satisfactorily unless you put your heart and soul into it. I could not do this job unless I worked at it. Some people treat it as a job. If you are flying with such people it becomes immediately evident because they do

> not know anything. They can do the bits they are required to do, but they do not know anything else. When I was flying as a Captain I would occasionally fly with First Officers or Engineers who were like that and really they were almost useless .... You do not do it for the money ... you are not a glorified taxi-driver. It is a very technical and demanding skill. To do it properly you have to be very conscientious. You do have to be an enthusiast to be good at the job. I think you can probably get away with it without being terribly enthusiastic ... but to do it well you have to be enthusiastic about what you are doing ... you have to understand your 'bit' of aviation ... this is not going to happen if you just turn up one-hour before your duty, do the paperwork and get on board. You have to do much more than that.

One interviewee was asked why he had given up his teaching career to become a pilot. He offered three reasons. First, the act of controlling an aircraft has an addictive quality. Secondly flying is a visceral experience. Thirdly, the industry has social utility:

> Flying is a bug. Once it has caught you it will never let you go. The beauty of flying is that at the end of the day the job is done. No thinking about what I have got to do tomorrow ... once the task is done, all your planning is done, the job is complete. You walk away .... There are only certain people who have the skill and ability to do it ... from that point of view therefore I suppose it is a vocation, because it is a skill available to only a certain proportion of the population .... There are wonderful pleasures, like the view ... looking out at con-trails ... being able to do that ... helping passengers ... getting them on their way ... trying to get your passengers to their destinations as safely as possible and on time ... giving them as smooth and pleasant a flight as you can ....

This respondent had become a commercial pilot after many years teaching and instructing (albeit in fields not unrelated to aviation). Having gained a BSc in Electronics and a Masters in Electrical Materials Science he worked as a teacher before joining British Aerospace (BAe) in Saudi Arabia where he taught avionics at the King Faisal Air Academy. On returning to Britain he taught at Solihull Technical College. He secured his first commercial flying job in 1986 at the age of 40. (He retired from flying in 2002 on medical grounds.)

As demonstrated earlier, pilots' career paths are far from predictable. While this pilot had moved from teaching to flying (at the age of 40), another had moved from flying to teaching, aged 44. The career change had provoked a major lifestyle change. As a trainee teacher he received a State bursary of about £6,000. Married with two young daughters he was planning to move into more affordable accommodation. This interviewee's career path seemed to mark him out as a vocational flier. Enthusiastic about flying from an early age he had joined the RAF. Over the course of his commercial flying career (which he had commenced in 1989) he had worked for eight airlines. The fact that most positions proved to be short-lived (six lasted no more than one year, and the longest lasted five years) and that he had lost confidence in airline management may have influenced his decision to leave the industry.

*Provenance: the role of the military*

A minority of respondents had flown for the military. Of those respondents born in the UK, forty-five percent had flown for the military (either the Army, Royal Navy or Royal Air Force). Of those born outside the UK a smaller number – twenty-four percent – had flown for the military. Given the gradual reduction in military spending by most Western nations it is reasonable to assume that the number of ex-military pilots in commercial aviation will reduce. The introduction of unmanned air vehicles (UAVs) into military aviation may also impact the number of pilots trained by the military (although it should be remembered that military flying has seen several false technological dawns, as with the 1950s prediction that manned interceptors would be replaced by long-range surface-to-air missiles like the Bloodhound. It was argued that such missiles would provide a cheaper and more reliable solution to the problem of intercepting Soviet bombers. The same rhetoric is being used today in relation to UAVs (although the military problem has changed to one of regional conflict and asymmetric warfare).)

## Summary

Commercial flying is clearly not the preserve of the sons and daughters of the rich, or of Ivy League graduates. While the majority of those surveyed came from comfortable backgrounds, a significant minority did not. Most had not attended university. The research showed that respondents' and interviewees' career paths could be highly convoluted.[2] This finding resonates with popular accounts of pilots' career paths. Consider, for example, the following life history reproduced from *99News*: 'In 1965, while studying computer programming at Orange Coast College, I was also teaching at an instrument ground school. One of my students ... hired me to fly company executives and clients in a Cessna 182 .... For the next 10 years, my corporate flying career was diversified and adventurous .... [However] the 1970s were not easy times for women in professional aviation and I chose to return to college for a career change to dental hygiene' (Krisbergs-Culp 2001).

The questionnaire survey and interviews revealed that some had come to flying late. Others who might have been expected to continue flying until they retired had left the industry to pursue a new interest (as with Ms Krisbergs-Culp, above). It would seem that commercial aircraft are flown by people from every type of background and with every level of educational qualification. Regarding pilots educated in the UK, for example, qualifications ranged from five O-levels to Masters degrees.

Some pilots described commercial flying as a job and some as a vocation. Although considered a testing and worthwhile occupation, flying was not felt to be entirely devoid of romance. As one pilot explained: 'There are wonderful pleasures,

2 One interviewee went so far as to claim: 'All careers are non-standard'.

like the view ...'. While not considered to be an intellectual activity,[3] it was felt that flying an aircraft safely and economically required 'skill and ability'. These findings broadly support Gant's (2003) analysis of pilots' motivations: 'People become pilots [because] they love flying, and they love using their talents and being respected for them. And mostly, they love the feeling of belonging to this strong family called aviation'.

---

3 An intellectual activity involves, amongst other things, the use of abstract reasoning. Heuristics or allegories may be used to play with or communicate complex ideas.

# Chapter 4

# Pilots' Lifestyle and Socio-economic Status

Most respondents and interviewees said they were in a relationship. A majority of respondents' partners were in paid employment. Of these, nearly thirty percent worked in commercial aviation, the majority of these working as flight attendants. It was noted that aviation-related occupations had a higher divorce rate than non-aviation occupations. The questionnaire sought to establish to what degree pilots who were divorced attributed their failed marriage to work patterns (that is, rostering). Sixty-three percent of divorcees who chose to answer this question claimed that rostering played some part.

As to why rostering patterns contributed to the break-down of their marriage, the following reasons were given: '[I was] not there to solve problems. Too much responsibility devolved to wife and too much temptation for both of us'; '[I spent] long periods overseas'; 'Long [periods] flying with attractive young women fed my wife's obsessive (and unjustified) jealousy'; '[There were] perpetual night-stops. [I was] always tired. [It was] difficult making time together worthwhile'; 'On promotion to Captain I was moved from BHX [Birmingham] to LGW [Gatwick]. Separation and long drives added to stress and communication breakdown. No assistance from company'; '[There was] constant weekend work [and] multiple roster changes'; '[My wife] had too much time on her own with two children ... could not cope with the uncertainty of roster-patterns'; 'Rotation into Nigeria was eight weeks on and four weeks off. Eight weeks is a long time. [I] spent half of the four weeks getting to know my wife again, the other half preparing to go [to Nigeria]'; '[I did] voluntary Search and Rescue flying with the Civil Air Patrol. My wife resented my not turning down missions. They were, of course, unscheduled'; '[Rostering played a part] by detaching me from the normal socio-economic week-based environment. Also the continual emphasis on productivity dictates that the early-to-late weekly roster given to UK charter pilots ensures unreal fatigue at the season's end'; '[I was] never at home during the day. [I had] few days off and was unable to take leave due to company commitments'; 'This is a great career, and I enjoy it immensely, but I had no idea about how the lifestyle would affect my personal life'; 'I have changed airline jobs in the past in order to ensure a family lifestyle'; 'You asked about failed marriages. I have never been married but I was engaged once and my career choice definitely played a part in us not getting married'.

One interviewee explained how a company rationalisation had necessitated that she live apart from her husband when on 'earlies':

> I had been based at East Midlands for four years when the airline decided to close EMA's Embraer base. We were allocated other bases around the country. I was moved to Manchester (MAN), which was the best choice for me, but it was still a drive [at this time the interviewee lived in Egginton, Derbyshire]. For an early start out of MAN I would rise at 03:45 and leave at 04:15. I'd get to MAN at about 05:15 ... the staff car park was a good twenty-minute bus ride away and I would get to the Crew Room at about 05:50 ... so it was a long way to drive before doing your four-sector day .... Consequently if I had a string of early starts I would stay with my partner's mother and father. They lived about thirty minutes west of Manchester. This was convenient, but I did not want to spend my married life living with my in-laws .... Regarding stand-byes, we pleaded with the airline to increase our stand-by report time[1] to ninety minutes. This would have meant that all of us from EMA who had been sent to MAN could have done our stand-bys from home. This would have made a massive difference to our quality of life. In the event we all developed our own ways of dealing with a seven-hour stand-by. I would drive to Keele University near Stoke-on-Trent and go to the gym for ninety minutes or so. Then I would sit in my car for six hours and read the books for the course I was doing. I might do that for three days consecutively. The rule at the time was that you had to be at work within sixty-minutes of receiving your telephone call. I am not 'Ms Goody-Goody' but I could not have coped with the stress of being in bed here [Egginton] at 05:00 and getting a telephone call asking me to be at MAN's Crew Room by 06:00 .... I could not have lived like that although some of my colleagues did ... I would have had to compress my usual drive in to forty-five minutes. In a sense I was my own worst enemy because I said all along that I am not going to get speeding fines because of my airline and I am not going to crash my car because of my airline. If the airline had increased the report time to ninety minutes then at least when I was on stand-by I could have been at home.

While staying with her husband's mother and father she chose not to take a room in their house: 'I lived in their caravan parked in their garden. There was room in their house but I wanted my own space'. Another pilot (also a female) also had first-hand knowledge of the caravan lifestyle:

> [I lived in] a caravan. It was thirty years old; covered in mould, inside and out, and had no water. There was a shower block two hundred metres away. It was horrific going down there in the Winter. No lights. After I moved out of the caravan it was blown up for target-practice. That was by far the worst temporary accommodation, happily .... The pros of living in digs are that you don't have to drive very far .... The cons are that you just never have a home ... you feel like a vagrant all the time. It is just horrible. If you are staying at someone else's house you are in the way. You cannot cook properly.

While it is clear that the aviation industry's dynamic character creates stresses and strains for its employees it is worth remembering that this is not unknown in other industries. Construction workers, for example, may also find themselves living away from home (and domiciled in caravans) for extended periods. During the UK

1 Given that pilots on 'stand-by' may be called in to work, they are required by their airline to live within a certain commute time of their base. This might be sixty minutes from being called at home.

recession of the early 1980s, many unemployed British construction workers found jobs in Germany. In the UK, the medical profession has long protested the long hours worked by junior doctors (who, like pilots, are responsible for the safety of members of the public, albeit on a much more limited scale). Having said this, it is clear that the vagaries of the flying profession can create significant pressures for pilots' partners or husbands/wives, as may be gathered from this retired pilot's response to the question: 'Did your wife ever discuss rostering with you?':

> Oh yes. The usual question was: 'Isn't there anyone else at work or are you the only one?' I have known people whose wife gave them absolute hell because they had to go in to work, even though it was their job. For the luckier pilots such relationships did not last, because it is bad enough being pressured at work without being pressured at home as well. There are lots of stories of people going out to dinner with their neighbours and falling asleep half way through the meal, or having to leave early, or being unable to drink. It could be very disruptive where relationships were concerned. It does not matter whether you fly scheduled for £150,000 per annum or fly for a freighting company, the amount of money you earn does not seem to mollify your partner, probably because they would rather have you at home.

This pilot commented on the temptations (as he saw it) of long-haul flying, and on the use of sympathetic rostering to reduce the strain on family life:

> Flying can be very wearing on families. I have friends who have had four wives. The divorce rate is quite high and there are a variety of reasons for it. The most obvious reason is that there is more opportunity to stray because you are working with a bunch of very attractive females the whole time and you get remarkably blasé about it. But occasionally people do fall for each other ... the opportunity is there ... you are miles away from home and nobody knows who you are ... this would not apply if you worked in the local bank ... people do become serial husbands .... Apparently there was an ex-British Airways Concorde captain who ended up flying small turboprops for CitiExpress because that was all he could get. He was in his sixties and all he had in life was a twenty-three year old girlfriend, a one-bedroomed flat in Crawley and a Ford Escort diesel van[2] ... because all the money had gone on every other wife and girlfriend ... that is how it can be ... people can lose everything. The wife and I attribute our long marriage to the fact that we never met anyone else at the same time we were pissed off [disaffected] ... you cannot be moralistic ... the opportunity is there, and if there is a problem in the relationship .... At my last carrier I used to get on with Crewing .... At one point my wife had breast cancer, then she had a lumpectomy .... But I needed to be at certain places at certain times. So I telephoned Crewing to make them aware of that. The chap in Crewing said: 'Fuck the carrier. If you need to go somewhere, call me'. So what happens then is that you will do the most extraordinary favour for the chap on your day off, because you know that he has looked after you in the past and would do so in the future. Rostering that you have a say in can make you feel a hell of a lot better .... Companies do not do enough of this .... When you joined Air Europe they used to make you meet the head of rostering, and make you walk through rostering to meet people because most of them are disembodied voices

2 A utilitarian vehicle.

> on the telephone .... When you know the people in rostering you can feel better about the whole job ....

Clearly a career in commercial aviation can create tensions and difficulties within relationships. Such difficulties need to be seen in a wider context, however, as with the slow decline in the number of marriages in the United Kingdom and higher divorce rate: 'Between 1971 and 1985 the proportion of families headed by married couples fell from 92 per cent to 86 per cent, while the share headed by lone mothers increased from 7 per cent to 12 per cent' (Hudson and Williams 1989, pp. 110–111). Taking a longer view, Lewis (1984, p. 221) notes that between 1960 and 1980 the UK saw a 400% increase in the divorce rate. In the UK in 2002, '... the number of divorces increased by 2.7% .... The average length of marriage for divorces granted in 2002 was 11.1 years ... ' (Carvel 2003). According to *Social Trends*: 'In 2003 the number of [UK] divorces increased by 4 per cent to 167,000, from 161,000 in 2002. This was the third successive annual rise' (Summerfield and Gill 2005, p. 24). A spokesperson for the marriage guidance agency Relate, notes: 'The social and economic pressures on marriage now are quite telling .... [Britons] work the longest hours in Europe' (Hall cited in Carvel 2003).

Questions about rostering could illicit strong responses, both positive and negative. Sometimes a pilot would speak of both good and – as he saw it – bad treatment by rostering, as with this testimony from a retired commuter pilot for a UK-registered carrier:

> [O]ne experience tempted me to give it up. I think it happens to everybody from time to time .... I was due to get married that afternoon ... I was flying in the morning: East Midlands to Aberdeen and back again. I arrived back at East Midlands and was just about to leave the aircraft when my mobile telephone rang: 'Ian can you stay on, please. The next crew is not available.' I blew up and did not do the flight. They found the other crew .... I got really annoyed .... [At] Brymon the policy seemed to me to be: 'Look after your crews and they will look after you'. You move heaven and earth ... let's get the service going ... let's get the passengers there ... if we do that they are likely to buy more tickets.

This same interviewee had also had good experiences (but with other carriers). Asked if he had ever told Crewing that he was 'too fatigued to fly' he replied:

> Yes, and they reacted in a very professional manner because I explained why I was too fatigued to do the duty ... it was during the winter and we had had a week of very bad weather, so flying was hard work, and I think I had had a roster change at the last minute ... I was flying the Short 330[3] for a small company where we were on first-name terms. I think the Chief Pilot called me and said: 'There's this trip. You can just about do it on hours. Are you prepared to?' Everybody wants to do the best for their company, as well as for themselves. But I just said: 'Look, I have had five days in the thick of it every night. I am sorry. Can I say no?' I telephoned the Managing Director and told him of my

3 The Short SD330 was a small, twin-turboprop freighter/airliner able to carry 30 passengers.

> decision. He said: 'Ian, don't worry, mate. It is perfectly alright. I fully appreciate the circumstances'.

Another interviewee described a duty that had prompted him to resign:

> I used to work for a long-haul charter carrier with bad rostering practices. On this occasion we had been rostered for a trip from London Gatwick to Antigua using a 767-300ER. The carrier had planned this flight using still-air wind conditions. The wind-component used on other trans-Atlantic flights was twenty knots headwind. That gave the airline an advantage because the flight seemed 'do-able'. I rose at 06:15 that day and commuted to work where I and my colleague calculated from the flight plan that due to this 'fiddle' – ignoring the wind – it was actually not possible to fly the 767 with 326 passengers direct to Antigua. We then had to consider whether we should cancel the flight or tech-stop in the US to pick up fuel. We actually stopped over in Bangor, Maine. We calculated that if we could refuel at Bangor in fifteen minutes we would land between two to three hours into the discretionary period. We then asked the Flight Operations Director for written permission to go up to three hours into discretion on this one occasion. We wanted some kind of legal authority from the company. As it happened we refuelled in Bangor in twenty minutes. We then negotiated a short-cut to Antigua which meant that we would make our destination two hours and forty-five minutes into discretion.[4] Bear in mind that if we were to land over three hours into discretion then that would have constituted a serious offence for the airline. So we were sailing fairly close to the wind. We were under a lot of pressure from the company. To avoid this situation we would have had to abandon 326 passengers at our tech-stop in the USA. So we headed down to Antigua. On the way we ran into a headwind. The Captain said he was too fatigued to do the approach and landing, so I did it. I flew a non-precision approach. I was pretty tired. We were going to land very close to three hours into discretion. The approach is an offset DME approach because the VOR sits on a hill to the right of the threshold. The approach must be done correctly. When the Captain called the decision-height I looked for the airfield but could not see it. There was a cloudburst. If we had done a go-around we might have had to make a forty-minute diversion. We would have gone well over our discretion limit. We might have had our licences endorsed. Fortunately the cloudburst passed, and I landed the aircraft. We were on-blocks two hours and fifty-seven minutes into our discretion. I was shaking. After that I resigned from the airline.

As to *how* company rostering departments could push some crews to the limit of their endurance, one interviewee offered this observation:

> The big problem is that the CAA lay down the absolute maximum and say that standard rostering must be less than this. It is to give you sufficient flexibility so that in an emergency you have room to utilise your crews more efficiently. We do not expect you to roster your crews to the full extent permitted by this all the time. But, of course, because it is legal, companies do it ... because the CAA do not have the resources to police the regulations

4 It would seem that in this case the airline viewed discretion as more of a target than a contingency (in the same way that some carriers have been accused of viewing the flight-time yearly maximum as more of a norm or target than a contingency to be used only in exceptional circumstances).

> as they should do ... it costs too much ... they spot-check, but airlines can get away with much more than they used to ... rostering officers consider it a challenge to consider men as machines.

Another interviewee discussed the impact of stand-by duties on family life:

> Yes [my wife discusses rostering] most days. She raises it because of the need to co-ordinate family life against a completely random work-pattern. She wanted to find out in advance when I was going to be too tired to be a useful participant in family life. My wife could look at the roster and predict what I would be like on certain days. She might say: 'Here comes a week of stand-byes. That is no use to anybody'. The issues change when the roster changes. All the time it is a discussion about how you are going to manage your life.

One pilot's response to the lifestyle questions smacked of exasperation: 'Once I have driven 1,200 miles a month (minimum), flown 85 hours (average) plus the rest of the duty-time, kept up with all the company paperwork and checks, done two lots of housework and the shopping, put in enough hours of Open University study (to be able to get a real job), kept up to date with Eastenders, oh and slept ..... where on earth do I have time for a relationship?' In her book *Women in England 1870–1950* Lewis (1984, p. 222) comments: 'It is still easier to do a demanding job if marriage and motherhood are renounced'. As for the 'Superwoman' ideal of the busy working mother she notes: 'The gains of the post-war period have tended to favour those women ... who had the means to buy childcare, domestic appliances and, often, domestic help' (Lewis 1984, p. 222). Clearly while women have made advances in the employment market, life is not particularly easy for self-actualisers on low, or even moderate incomes.

As revealed earlier in the book, some consider aviation to be a glamorous industry. Judging by the following statements made by two of the six interviewees this is not exactly true in all circumstances. In the first statement, a retired pilot recalls how he ended his flying career. In the second, another retiree recalls his time at a UK-registered regional carrier where he flew turboprops:

> Towards the end of my career I was doing short-haul, unfortunately. I preferred long-haul. I was on the Airbus. For me getting up at 04:00 or staying up all night and finishing up at the place you started from was not the most interesting type of flying. For me the most interesting type of flying is scheduled long-haul, because you feel you are taking people somewhere they want to go ... it is much more exciting taking people to, say, New York [during his commercial flying career the interviewee had lived in New York for two periods]. You know it is going to be a better experience than going somewhere else and also it gives you the chance – even if you are only there for twenty hours – to get off the aircraft, eat a meal in a different place, have a few drinks with the crew and come back the next night. For some peculiar reason that was much more satisfying to me than going there and back in the same day. When I was flying long-haul I never had a situation where I would get back to base and could not remember where I had been. But this was a regular occurrence when I was doing short-haul. As you walked in to the Crew Room people would ask: 'Where have you been?' and I would say: 'Yes ... where

> *have* we been?' Usually you would have to consult the paperwork, because by the time you got to the Crew Room you had wiped it. This happened all the time. It was another strip of concrete, another turnaround, another return. However, had you gone down town for a meal at your destination it would be more fixed in your mind. It is rather like going down the motorway to a service station, picking up fuel and then coming straight back. Alternatively you can go down the same motorway and continue on to, say, Glasgow, and then have a night out.
>
> When I flew for Eastern out of Aberdeen we had to rest in the terminal with all the passengers around us. [I did not feel adequately rested] but then I was on duty the whole time, so I could not complain, really.[5] We did have a Crew Room at East Midlands.

Commercial aviation is a unionised industry (although, as noted earlier, not all airlines recognise trades unions for the purpose of wage bargaining). Asked whether he had found his union of help in disagreements over rostering and other matters, one interviewee replied:

> Generally I was not satisfied with BALPA. BALPA is a curate's egg. It is good in parts. It is an intensely political organisation with different vested interests. I am sure there were little feuds being carried on between people. The central issue, however, is that BALPA has to ensure its own survival, so it has to protect its own third-party relationship with the airlines. So the union will go so far on behalf of its members, then invariably it will back down .... BALPA seems to me to be a weak trade union with a difficult job to do. It seems that the only sanction a trade union has is to take strike action, but in the case of the piloting fraternity, which is small and close-knit, a strike would be very damaging for the individuals concerned. I think the strike at Cathay Pacific bore that out. Pilots were dismissed and found it difficult to work again ... so the ultimate sanction of going out on strike is not really available in practical terms.

Asked what use he had made of his trade union in rostering disputes, another interviewee replied:

> In the minds of many, BALPA are a very expensive luxury. Although as far as the press are concerned BALPA are the mouthpiece of UK pilots, there is a lot of criticism of the service they provide and what they charge. As a result of all that I have been interested in independent associations. By the way, BALPA stands for British Airways Pilots' Association .... I am sure they are doing a good job but they have never done a good job for me.

Although respondents and interviewees were not asked any direct questions concerning their personal income, they were asked to value their real estate. A random sample of 34 responses was analysed. The value of real-estate owned averaged out at £372,647 (at 2003/2004 prices) – over twice the average price of a house in the UK. Investing in real estate is one way of conveying accumulated wealth from one

5 The airline preferred to keep its crews on duty rather than give them a 'split duty', which would have allowed them private, off-duty time to relax and refresh.

generation to the next. While the figure of £372,647 might suggest that pilots are doing rather well, it should be remembered that property prices in the vicinity of airports may be higher than those elsewhere. Consequently, those who have to live in the vicinity of an airport get less for every pound they spend. It should also be noted that the value of real estate owned provides only a crude guide to career success and financial liquidity. It is possible, for example, that because of an inheritance a relatively poorly paid First Officer (or, indeed, Captain) could own real estate valued at many hundreds of thousands of pounds. Given the cost of taxing, maintaining, heating and lighting such a property, the incumbent could actually find her/himself with a relatively low disposable income (to cover day-to-day expenditures (e.g. food and travel) and long-term commitments like school or university fees).

Regarding family size, nearly two-thirds of respondents had at least one child. (In Great Britain in 1971 the average household size was 2.9 persons. By 2004 it was 2.4 (Summerfield and Gill 2005, p.20).) As societies develop and become more prosperous, the percentage of young people going to university tends to increase. In the UK, the number of full-time undergraduates increased from 414,000 in 1970/71 to 1,179,000 in 2002/03. Summerfield and Gill (2005, p. 36) observe: 'There have ... been substantial increases in the number of students in higher education over the last 30 or so years .... There were six and a half times as many female students in [UK] higher education in 2002/03 as in 1970/71 [and] two and a half times as many male students'. Three quarters of respondents with children of university age were putting at least one through university. Clearly, the majority of respondents were determined to give their offspring the best possible life-chance. Summerfield and Gill (2005, p. 36) note: 'Participation rates [in UK higher education] for the non-manual social classes increased from 27 per cent to 50 per cent [between 1960 and 2001]'. Over the same period participation rates for the manual social classes increased from four percent to nineteen percent.

Access to higher education facilitates access to the best jobs. By putting their offspring through university respondents were – either by accident or design – going to reproduce their own social, economic and professional status. As Hudson and Williams (1989, p. 63) explain: '[I]nequalities in access to ... the educational system both reflect and help reproduce class structures within the UK. Undoubtedly, access to "top jobs" in UK society is closely constrained by class and educational background. Class structure and education are mutually reinforcing'. It has been noted that students from comfortably-off families (and pilots' families generally fall into this category) tend to do better at university than those from less well-off families: 'University life [in the UK] is increasingly becoming a two-tier experience, with wealthier students studying hard and taking CV-enhancing work experience while their poorer classmates work long hours in low-paid jobs, which means their grades suffer as a result .... [O]ne of the researchers on the project [that examined student lifestyles] ... said: "Students' experiences at university are getting increasingly polarised. So you have higher income students who don't need to work in term time and do interesting work in holidays to enhance their CVs. Then there are those who have to work in term time and whose academic work suffers. The work they do is for short-term cash benefit rather than long-term career benefit"' (Curtis 2005).

Few respondents smoked. Of those who did well over half were in their 40s and 50s, over half smoked less than ten cigarettes per day,[6] and most took exercise (although definitions of exercise varied). Summerfield and Gill (2005, p. 102) note: 'Over the past 30 years there has been a substantial decline in the proportion of adults ... in Great Britain who smoke cigarettes, from 45 per cent in 1974 to 26 per cent in 2003/04'. One retired pilot ventured that in the last ten years he had met only one colleague who smoked. Asked whether smokers had ever had a problem with not being allowed to smoke on duty he said: 'No, because they ignored the regulations and smoked until they were told not to'.

Overall, nine out of ten respondents took some form of exercise. Summerfield and Gill (2005, p.177) note that, in 2002/2003, fifty-nine percent of Britons aged 16 and over participated in a sport, game or physical activity (including walks of two miles or more). One interviewee described pilots' work environment as inherently unhealthy:

> The job is not conducive to good physical health. You gain weight. I have gained at least two stones since I started flying, which is awful. Especially for women. I am very aware of the fact that I might be getting deep-vein thromboses. That really does concern me. The backs of my legs get really stiff. It is bad for your back. A lot of people have back problems. The seats are not comfortable. They are not very supportive. You can adjust them but the support is just not in the right place, especially on the older seats. The seats on our Boeing 757s are twenty years old. It is like sitting on a board. It is rubbish. Also being in an air-conditioned environment for that length of time ... it dries your skin out ... it's horrible. I don't know why we do it really!

Harris (2005, p. 209) notes: 'Studies of pilots' drinking and flying behaviour are relatively sparse compared to the wealth of literature available from the study of drinking and driving'. Over ninety percent of questionnaire respondents said they consumed alcohol.[7] Of these a small number had discussed their alcohol consumption with a qualified medical practitioner. Summerfield and Gill (2005, p. 101) make this observation on alcohol consumption in the UK: 'Alcohol consumption is higher in managerial and professional households than in routine and manual ones. In 2003/04, 76 per cent of adults ... in managerial and professional households had had a drink in the previous week. This compares with 59 per cent ... of adults in routine and manual households. The proportion of people in managerial and professional households drinking on five or more days in the previous week was twice that of adults in the routine and manual group (24 per cent compared with 12 per cent)'.

The question of alcohol consumption amongst UK pilots hit the headlines in October 2000 following a TV documentary on the subject. According to Harris (2005,

---

6 *Social Trends* defines a moderate smoker as someone who smokes between 10 and 19 cigarettes per day (Summerfield and Gill 2005, p. 102).

7 For the purpose of this study alcohol, like nicotine, is classed as an addictive drug.

p. 200) the *Dispatches* programme revealed '... some interesting attitudes toward the use of alcohol [amongst] crews. Of particular note was one opinion expressed that even in the event of a hangover when flying the following morning, everything would be "OK" and safety would not be compromised as the automated systems would look after the aircraft during critical phases of flight'. Of course there is no suggestion that any of the respondents and interviewees who said they consumed alcohol violated the various workplace rules and regulations pertaining to the drug. The figure may, however, be taken to indicate a wider set of issues. Harris (2005, p. 206) notes:

> Canada Market Research (1990), Ross and Ross (1995) and Maxwell and Harris (1997) ... all reported that job-related stresses were frequently cited as a common cause of heavy drinking in professional pilots. These stressors included being away from home, job-related fatigue, long hours and boredom .... Long-haul flight crewmembers commonly reported drinking socially with other crewmembers after flights and ... also as a means of helping them to relax, especially after crossing several time zones (Sloane and Cooper 1984).

Author and pilot of 30 years standing, Laurie Taylor's (cited in Prince 1990, pp. 106–107) recollections provide a social backcloth to alcohol consumption amongst pilots:

> There are those [pilots] who turn to alcohol as a relaxant ... while others look to it as a stimulant to keep them awake .... Domestic disruptions and a blitzed social life are other influences. Where there is that kind of highly-strung and taut lifestyle ... there are going to be occasions when somebody snaps .... After a long flight, all pilots are wound up like a coiled spring: I always was. It's impossible to go straight to bed and expect to sleep; you're wasting your time. I would always have a couple of beers after every flight, whatever the time of day, even if everyone else in the hotel was having breakfast .... What we did was healthy. Our drinking was out in the open, and was, without a doubt, therapeutic.

Regardless of its perceived therapeutic role, the social and economic costs of alcohol misuse are significant. Surveying the American experience of misuse Bartol and Martin (1998, p. 340) observe: '[D]rug and alcohol abuse ... typically leads to increases in absenteeism, workplace accidents, and use of medical benefits, and causes declines in productivity [sic]. A recent report indicates that productivity losses related to alcohol problems amount to about $86 billion per year ... '.

Most respondents owned at least one car and drove to work, the average journey time being 36 minutes. Obviously the time taken to commute to and from work added to the length of a pilot's day and had implications for fatigue and stress. As one interviewee whose longest commute had been a car journey of up to two hours observed:

> I did not like commuting such a long way because it made the days extremely long and very tiring. It was in the summer so it was boiling hot. I was jaded when I arrived at work but exhausted when I got home. I did not have the option of staying up there at the time

> because my airline told me I was going to be moving again, so I decided to stay put and commute.

Another pilot (a respondent) ventured that as UK FTLs were formulated in the days before 'gridlock', they required urgent revision:

> In the south east of England current house prices mean that pilots often commute long distances to travel to work. Poor car-parking facilities for crews result in some of my pilot-colleagues having four hours added on to their working days, just spent travelling to and from work.[8] Fatigue inevitably affects performance. The airline is unsympathetic to those who cannot live closer to the airport. CAA FTLs were devised in the 1950s when there were fewer cars on the roads and airline pilots were paid larger salaries in real terms than today. Current FTLs take no account of the change in these factors and the increasing commuting times pilots must endure, coupled with where they can afford to live.

The number of cars on the road grows year on year. According to Summerfield and Gill (2005, p. 166), between 1971 and 2002 the percentage of UK households with no car fell from 48 per cent to 26 per cent. They point out that with more cars making more journeys average traffic speeds tend to decline. Between 1993 and 2003 average traffic flows increased by seventeen percent (Summerfield and Gill 2005, p. 167).

## Summary

While most respondents claimed to be in a relationship, there was comment to the effect that the pilot's lifestyle could create problems for partners. Some pilots worked long hours. All pilots worked in a risk-laden environment. Fatigue and stress are not conducive to a happy home life. In order to get or keep a job some pilots were obliged to live away from home for periods of time. Forced separation had the potential to create tensions within relationships and to impact pilots' self-esteem. While many pilots made an effort to keep fit, most drank. It is possible that some consumed alcohol to counter the effects of stress, alleviate boredom or numb feelings of loneliness.

Pilots had a mixture of good and bad rostering experiences. Night-time operations could be especially fatiguing, as could extended stand-by duties. Duties could put a strain on relationships. It was felt that much could be gained if rostering departments tried to match duties to pilots' preferences. Some pilots tried to cultivate a relationship with their rostering department. Moving to a new base could produce extended commutes, thereby exacerbating fatigue problems. It is commonly agreed that fatigue impacts performance: 'As fatigue accumulates, performance degrades .... Involuntary "microsleeps" begin to occur .... Attention lapses occur, and working memory errors increase .... Problem-solving skills erode, nonessential activities start

8 At some UK airports (Gatwick, for example) flight crew parking is located at some considerable distance from the crew rooms, necessitating a time-consuming bus journey.

to be neglected and vigilance tasks become increasingly difficult to perform [a major issue as pilots today spend much time monitoring advanced automated systems] .... Activities judged to be of lower priority tend to become more neglected, situational awareness suffers and the risk of errors and accidents increases' (*Air Safety Week* 2004a, p. 3).

Flying, of course, had its financial compensations. Pilots with established careers could afford a reasonable lifestyle. They could afford to help their offspring. They could afford reasonable accommodation. They could afford one or more cars (one respondent owned five cars). The issue for successful pilots, however, was how they could create the time and space to enjoy their material rewards and maintain high quality personal relationships.

## Chapter 5

# Pilots' Commercial Flying Careers and Perceptions of the Industry

While there was no stereotypical recruit to commercial aviation, almost forty percent of respondents had been military pilots (flying either fixed or rotary wing aircraft for either the army, navy or air force). Amongst the over sixty percent of respondents who had come to flying via the civilian route, one felt strongly enough about civilian flying training to add the following to his questionnaire:

> The road to an airline cockpit is taxing. Most flight schools are interested only in your money and not your progress. They won't, for obvious reasons, give you the straight dope on how hard the road ahead is. It is hard to make people who are not in the industry understand how many hoops you had to jump through. Why you took a $5-an-hour job pumping gas at an airpark in 120-degree Phoenix heat so you can hopefully get the chance to fly cargo two days a week. And why you had to quit your previous lucrative position as Assistant Chief Pilot of a large flight academy to pump that gas because being a flight instructor only looks good on a résumé for so long.

This respondent's statement illustrates not only that the path to becoming a pilot can be exceedingly tortuous but also that to travel that path, would-be pilots must remain absolutely focused on their goal. They must be single-minded, dogged and determined. They must be willing to tolerate short-term pain for (hoped-for) long-term gain. This view of pilot training resonates with that of a British Airways Boeing 737 First Officer whose testimony BA reproduced on its training bulletin board *Explore Our Working World*: 'Its an extremely tough course ... [you] really need to love flying and have a motivation .... It's important to have a back-up plan of some sort' (cited in British Airways 2005).

It was rare for a pilot to stay at the same carrier for her/his entire career. (It is important to note that while some elected to move, others were made redundant or were employed on short-term contracts.) About a third of pilots stayed for between one and 12 months at a carrier. Summerfield and Gill (2005, pp. 58–59) make the following observations on labour mobility in the United Kingdom: 'In spring 2004, one in ten employees of working age had been in their current job less than six months, and a similar proportion [10%] had been in the same job for 20 years or more. Most people (80 per cent) had been in the job for a year or more and around one quarter (23 per cent) had been in their job between two to five years'.

Many airlines use their more able pilots in non-flying roles, like recruitment, safety monitoring and ground-school. One third of respondents performed additional, non-flying duties. Many were involved in training which they said they found rewarding. (Some pilots said they needed greater stimulation. One respondent commented: '[F]lying four sectors a day ... five days each week is very ... repetitive. Not much scope for ... creativity ... '.[1]) Worryingly, only fifty-nine percent judged that they were allocated sufficient time to perform their additional ground-based duties *to their satisfaction.*

Sixty-five percent of respondents said they would *not* encourage their children to become a commercial pilot. Those who said they would encourage their offspring to become a pilot cited several attractions, including income, variety and opportunities for 'personal growth' (self-actualisation). Those who said they were ambivalent or who said they would not encourage their offspring to become a commercial pilot cited what they saw as the industry's ruthlessness, lack of job security and unsympathetic management. Other deterrent factors included the high cost of training, the make-or-break character of recurrent medical assessments and competency tests (like simulator checks and check rides), the perception that pilots were being de-skilled and the job's impact on physical health and relationships. Regarding the perceived deskilling of flight crew, one respondent observed: '[The job is] not sufficiently ... challenging for a bright graduate candidate [and] uses too narrow a range of skills in too repetitive an environment'. Another commented: 'My children are too ... academic to really fit in as pilots'. Several respondents took the opportunity to provide a more detailed commentary on their industry. Their perceptions or constructions of what can seem a glamorous occupation are noteworthy:

> The days of the pilot being a respected member of the community who retained special skills that were admired by the traveller have, in the main, gone, and a lot of pilots have become unionised train drivers. Most pilots today seem to want more payment for less work. The aircraft have become so technologically advanced that the pilot has become a computer-operator and the traditional pilot skills are diminishing. The level of regulation has become a real burden for the pilot, who has continually to learn new rules and new equipment. The safety of airlines has undoubtedly benefited from the new equipment, but when it goes wrong I sometimes worry about the outcome.

> In thirty-five years civil aviation has 'dumbed-down', like much of society. To be selected for training in the 1960s in the UK you needed a middle-class background with good academic qualifications. Now it is open to all, which has taken the halo-effect from the profession. Like train drivers in the Nineteenth Century and ship's captains in the late Twentieth Century, pilots are now reduced in social status and are seen as just another worker. The required skill-level has been reduced to allow cheaper workers to fly aeroplanes. The job's technical complexity has become much greater, but pilots are trained just to the minimum level, and the complexity is kept from them.

1 This statement supports the view that flying is not an intellectual pursuit. Rather it is a craft that requires various competencies and skills, like the ability to perform complex mathematical calculations while flying an aircraft.

> The job has become less fun in the last few years. I am not sure if the cause is economic, regulatory (security) or both. Commuting has become more difficult. And the travel benefits are nearly worthless due to high loads. I envision many pilots will leave the career field due to quality-of-life issues. Other would-be pilots might choose different jobs as well. This may lead to a shortage of pilots.

Interestingly, several of those who were critical of the level of financial reward owned high-value real estate. The respondent who stated: 'Being a ... pilot is not as lucrative as other professions' owned two houses with a combined market value of £680,000 (at 2004 prices). The respondent who said: 'The pay is now not so good' owned a five-bedroomed house valued at £700,000, a four-bedroomed second home abroad valued at $395,000 and three cars.[2]

While pilot salaries of £100,000 per annum in the UK and $110,000 in the US are not unknown, most pilots earn significantly less. In 2005, easyJet, for example, paid its senior FOs a salary of £39,000 and its junior FOs a salary of £32,000. By January 2006 the company was paying its senior FOs £40,584 and its junior FOs £33,060. (Similar variations obtain in the United States, where top-earners can gross roughly three times the salary of those at the bottom of the pay ladder.) Using easyJet's 2005 data for First Officers' salaries, a rough comparison can be made with other trades and professions. While First Officers were paid between £32,000 and £39,000, hospital nurses were paid £26,000, journalists £28,500, train drivers £32,000 and plumbers £35,000 (Triggle 2005). Clearly those who sit in the right-hand seat are not going to make their fortune flying commercial aircraft. It should also be noted that airlines are under great pressure to control costs – including pilots' salaries.

Regarding salary levels in the UK, Summerfield and Gill (2005, p. 70) note: 'In April 2004, directors and chief executives of major organisations, with median earnings of £1,791 [gross] per week, topped the earnings league for full-time employees .... Senior officials in national government, medical practitioners and aircraft pilots and flight engineers also had median weekly earnings exceeding £1,000 [actually £1,094]. The lowest paid of all adult full-time employees were leisure and theme park attendants and floral arrangers and florists, with median earnings below £200 per week, only just over a tenth of the earnings of the highest paid occupational group'. The median gross weekly pay for a judge, lawyer or solicitor was £832 and that for a medical practitioner was £1,168. Regarding these last figures it should be remembered that medians do not reflect the distribution of salaries. Given this fact, many solicitors, for example, would earn much less than the government-calculated median income of £832.

Those respondents who said they would not recommend a career in commercial aviation considered the established professions to offer the best prospects. Favourites were medicine, dentistry, the law, civil engineering and architecture. Also mentioned were the armed services, police service, information technology, telecommunications and the civil service (that is, civilian government service). A few suggested self-

2 In the UK in December, 2004, the average price of a residential property was over £152,000.

employment. Summerfield and Gill (2005, p. 70) note that, in April 2004, the median gross weekly pay for 'IT strategy and planning professionals' was £797, and that for 'Information and communication technology managers' was £795. It is clear that interviewees and respondents wanted their offspring to have a career, preferably in a venerable profession (like the law). Unlike commercial aviation the professions provide relatively stable employment.

*The military-to-commercial transition*

As mentioned above, almost forty percent of commercial pilots in the survey were ex-military fliers. Two of the six pilots subjected to detailed interview were ex-RAF pilots. Neither had found the transition from military to commercial flying easy. Difficulties encountered included the difference *in essence* between military and commercial flying, a perception that career commercial pilots were somewhat resentful of them, and a feeling that the ex-military pilot could hold a somewhat naive view of airline management. Regarding the essential difference between military and commercial flying, one of the ex-RAF pilots noted:

> It is a totally different world .... If the RAF went to somewhere like Tehran [before the Islamic Revolution a joint military-commercial airfield] we were struggling with their procedures and they were struggling to deal with us .... Transitioning is not easy. You still have to do the full ATPL. Most have more than one shot at the instrument-rating. So it is quite a hurdle. And of course you have to put money into it. Many, many military pilots would like to be civil pilots, but for one reason or another they do not make it.

Regarding the question of antipathy or resentment, the same interviewee commented:

> There was almost a contemporary of mine who joined a matter of weeks before me who had never been in the RAF and he permanently resented the fact [that I had] ... silver spoon and that sort of thing, or so he thought. RAF training is superb. It trains you to fly aeroplanes, although it does not teach you too much about procedures. He hadn't had that, and this became an issue subsequently when we were getting round to promotion. I had joined with an ATPL. I didn't mess around with the CPL. He held only the CPL. So when it came to promotion I was promoted first, which I think was correct. He didn't think that this was correct. There were few promotions. That was and remains a sore point with him, even though he is now retired.

Asked whether he had encountered resentment from anyone else, the interviewee responded: 'Oh yes. I would say that that is quite general. There is a sense in which RAF pilots are considered to be spoon-fed whereas fellows who started flying when they were twenty-two or twenty-three and had to go through the grind of flying schools resent it .... There are not so many RAF pilots around now, of course, but for most of my career, say one-half of us have been RAF-trained originally. I have frequently had remarks directed at me, particularly as the number of RAF-trained pilots diminishes'.

Asked whether, in retrospect, he wished he had managed his career differently, he stated: 'In retrospect it would have been far better if I had stayed in the RAF. Having decided to leave the RAF it would have been far better if I had gone to BEA and tried for a career there. I have enjoyed my life hugely since I moved on from the Ford Motor Company. I flew for Ford for about 20 years and I did not really like it. I left Ford at the end of 1988. I have enjoyed the last 15 years far more than the previous 20 years, partly because I have done all sorts of different things. Since I joined my present employer I have flown three different types, the Lockheed Electra, Boeing 727 and Airbus A300'.

Asked if he felt there were any difficulties in transitioning from military to commercial aviation, the second ex-RAF pilot-interviewee said: 'Military pilots who transfer have too much trust in management. They assume that airline managers are pointing in the same direction they are .... It takes about six months but they soon get the hang of it ... that it is commercial aviation and that a profit has to be made by everyone in commercial aviation .... In my day, the RAF did not worry about cost'. Despite the issue of not immediately understanding the commercial aviation manager's rationale, this interviewee claimed to have enjoyed the commercial world's greater 'social mix':

> Mixing military and civilian pilots is a good thing. In the RAF we tended to have the same type of personality. For example, one day three of us ex-military pilots went shopping somewhere in Texas. When we arrived at the check-out we had all bought the same pair of shoes and the same pair of trousers. We looked at each other and said: 'I don't bloody believe it!' The only thing that varied was the size. If I had had a couple of civilian colleagues with me one might have bought a skirt or something, I don't know, but it would have been different. Of course, during the Second World War you did not have just a single type of personality. All sorts were thrown together. But this changed. I think civilian crewmembers add to the range of personalities. I think everyone could have been in the military anyway ... stroke of the pen .... If you have someone with a bit of a personality problem they are more likely to be straight-through civilians.

Another ex-RAF pilot, actually a respondent, noted the commercial world's more enlightened attitude to such safety-enhancing protocols as Crew Resource Management (CRM):

> The RAF is miles behind the thinking in civil aviation with regard to CRM. Regrettably many senior people have a false idea of what CRM represents. Initial forays into this world have lacked top-level support, and have consequently been doomed. The latest push is balanced on a knife-edge and the Services' lack of enthusiasm for this valuable skill is one of the main reasons why I am retiring. The other concerns the imminent cut-backs of large numbers of skilled people, allied to the widespread poor personnel management, some of which is lamentable [this ex-RAF Squadron Leader had joined a UK-registered LCC in 2004].

Clearly, therefore, there are major differences between military and commercial aviation – differences that embrace both the operational and social aspects of flying.

## Summary

While most respondents constructed the occupation in a positive light there were some dissenters, as with the pilot who commented: 'I do not get any personal rewards any more. It is just a job that I am glad to go home and forget about'. Another stated: 'It used to be camaraderie, a sense of achievement, etc., .... Pilots and beancounters [accountants] do not mix and the beancounters have taken over the industry'. The most frequently cited plus points or rewards were: a sense of achievement and pride; friendship and teamwork (one respondent commented: 'I work with a pleasant bunch of people', while another stated: 'I get ... a social outing'); monetary reward (one pilot observed: 'We all have to pay the mortgage'; fun (one respondent stated: 'I am doing the best job in the world'). Similar opinions have been expressed elsewhere, as may be gathered from this Alaska-based corporate helicopter pilot's testimony (reproduced in *99News*): 'We work a 14-day hitch – 12 hours a day – and have 14 days off .... Our uniforms are of necessity and not of fashion. We don a flight suit and a bright orange survival suit that is very warm .... It might not seem glamorous. We usually go home smelling of jet fuel .... [but] we get to do it in a helicopter in what locals like to call "God's country"' (Saettone 2001).

Some believed that the job had changed *in essence*. Respondents described themselves variously as 'unionised train drivers', 'computer-operators' and 'just another worker'. There was a perception that the intercession of computers had diminished the importance of airmanship (the ability to understand and *instinctively* pilot an aircraft). One pilot suspected an economic motive: 'The required skill-level has been reduced to allow cheaper workers to fly aeroplanes'. There was a perception that the job had become more bureaucratic and that those seeking to join an interesting and valued profession would look elsewhere.

Respondents envisaged a dynamic future for commercial aviation (although it was felt that this dynamism could manifest as either innovation and growth or consolidation and collapse). One pilot commented: 'I foresee consolidation of the majors [and] big layoffs'. Another claimed: '"Dinosaur" carriers will not survive'. A third said: 'Boom and bust. Boom and bust'. While some consider LCCs to represent the future of commercial aviation,[3] respondents were not so sure. One predicted:

3 Some argue that the LCC model can be applied to both short-haul *and* long-haul operations. New York-based LCC jetBlue operates some very long trans-continental sectors. One of its services connects New York on the Atlantic coast with Long Beach on the Pacific coast. Nicholls (2006, p. 30) comments: 'JetBlue's route map clearly shows it has expanded well beyond its initial territory in the eastern USA'. Ionides (2006, p. 46) notes: 'The concept of medium/long-haul low-cost travel may still be ahead of its time, but a number of Asian carriers are willing to give it a try'.

'LCCs will price themselves out of the market'. Another foresaw: '[P]ublic disenchantment with the low-cost model'. A third said: 'More LCCs will start then fail'. At least one respondent predicted a more polarised industry with bare-bones LCCs and full-cost, full-service carriers operating side-by-side.[4]

Other predictions for the industry included: The 'saturation' of ATC; the advent of free-flight with ATC relegated to a watchkeeper role; more cargo operations; the development of more fuel-efficient aircraft; more congestion in the air and on the ground; government legislation to favour railways for short journeys; a more intense management focus on costs and productivity; a pilot shortage; the use of cheap labour from eastern Europe; more liberal flight and duty time limits; the deskilling of pilots; and a heightened security threat to commercial aviation. The pessimism of several of these predictions is noteworthy.

4 Some predict the advent of exclusive 'business only' widebody services across the North Atlantic, the continued expansion of tailored (but expensive) corporate flying and the development of supersonic business jets.

## Chapter 6

# Pilots' Concerns and Professional Culture

Interviewees and respondents had two major concerns. The first was the way in which commercial pressures impacted airlines' and regulators' behaviour. The second was a feeling that airline managements had become dangerously aloof. Regarding the first worry, the following comments were typical:

> The industry is out of control. Publicly the image is not that bad but deep down, airlines do not give a damn about anything (including safety) other than the bottom line. Example? When SQ006 crashed on take off (from the wrong runway) at Taipei a few years ago,[1] the CEO of Singapore Airlines held a televised press conference. What were his first words? 'Our shareholders have nothing to fear. We will do whatever it takes to protect the share price'. Almost one hundred people perished in that accident and his first concern was the fucking share price! And his attitude is typical.

> I am concerned about the degradation of working conditions and contracts while our requirements continue to increase.

> Airlines must use FTLs as a guide – not a target.

> The rostering and crew-control is appalling at [my UK-registered LCC]. If they could sort this out, that would be all they needed to do to retain a vast number of pilots and crew, and halt the periodic exodus from the airline – currently to BA, Virgin, easyJet, Cathay, Emirates, Qatar.

> A magic wand is required to give pilots some control over their lives. We all feel like slaves to our schedulers.

> The CAA seems unwilling to step in to stop rostering practices that lead to severe fatigue. They are supposed to be the industry's safety regulator, but appear to be steered by the commercial organisations themselves.

> Commercial aviation is a front-line industry in the liberalised economy. Regulatory bodies exhibit weakness in resisting the financial and industrial lobbies and therefore do not develop and enforce appropriate regulations.

---

1 SIA flight SQ006 attempted to take off from a closed runway that was under maintenance. The 747-400 hit construction equipment some four seconds after the First Officer called V1. Visibility was 500–600 metres. Over 80 passengers and crew were killed (Hannon and Kingsley-Jones 2000).

There was a feeling that commercial pressures might be impacting safety margins during turnaround. One interviewee remarked: 'Turnarounds were definitely an issue at my LCC. When I flew with them they were doing thirty-minute turnarounds. Subsequently I have noticed they have reduced it to twenty minutes. I do not know how they do it. Eating, lunches and perhaps you want to buy something with security in the way ... plus the fact that you have to do a certain amount of preparation when you are a two-pilot crew ... refuelling ... it is very easy for something to be overlooked when you are under pressure ... I would not like to go back to that sort of flying'.

Another interviewee claimed that commercial pressures created a 'hidden culture of short-cuts': 'Turnaround issues revolve around safety and blame. Nobody minds working fast to get the job done, but there can be perceived pressure to take short-cuts .... It is as much about perceived pressure than of actual pressure. The airline will be very sure to leave a good audit trail of its procedures to show on paper that everything is being done correctly ... but it is not the projected culture that counts, it is the hidden culture ... the whispers between pilots ... discussions over a coffee ... chats during turnarounds. A good example was that I was pressured by my Fleet Manager to not put items in the Technical Log until the last flight of the day because they did not want technical delays, which I thought was dreadful .... That same pressure had also been applied to other crews .... I was told by my Fleet Manager: "You are relatively new as a Captain. What you must realise is that we do not want delays around Europe. We want things repaired overnight". This was a very subtle way of saying: "Play the game our way". But the implications for flight safety are tremendous'. It is illuminating to consider interviewees' and respondents' concerns about ramp safety in light of regulators' warnings. The UK's Health and Safety Executive (HSE) has remarked: 'The apron is a busy and complex workplace. Workers face many hazards .... Currently, accident rates in the industry are well above the national average for all industries, and for ground handling and airport workers, accident rates exceed even those of the construction industry and agricultural sector' (HSE 2000, p. 1).

In 2006, the press reported on an investigation conducted by the UK's Air Accident Investigation Branch (AAIB) into maintenance lapses at British Airways. Like the interviewee quoted above, the Board focused on cultural failures. Following its investigation into a maintenance lapse that produced roll-control problems on a BA Boeing 757 aircraft, the AAIB noted: 'Ineffective supervision of maintenance staff had allowed working practices to develop that had compromised the level of airworthiness control and had become accepted as the "norm". Maintenance errors were not the result of wilful negligence, or any desire to perform a less than satisfactory job, but the result of a combination of systemic issues that had increased the probability of an error being committed' (AAIB cited in Walsh *et al.* 2006). It is worth asking to what extent commercial pressures induced the cultural failure identified by the AAIB. Walsh *et al.* (2006) note: 'BA employs 6,000 engineers, compared with 9,500 in 1995 but its fleet remains at a similar size as 10 years ago'. In recent years BA has undertaken a major cost-cutting drive. Baker (2006, p. 14) notes: 'BA's record on cutting labour costs has been draconian, with ... Rod

Eddington axing 14,000 jobs during his five-year tenure ...'. At the end of 2005 BA's new CEO pledged to re-invigorate this drive (Kane 2005). Responding to the AAIB, BA's safety head was contrite: 'British Airways prides itself on safety and recognises that we are always ready to learn from incidents and encourage open transparent reporting' (Young cited in Walsh *et al.* 2006). It remains to be seen whether this top-level commitment to isomorphic learning[2] is followed through.

Other respondents believed that commercial pressures might impact pilots' short and long-term health (both physical and mental):

> Airlines like SAS schedule crew with a change of aircraft and cabin crew several times a day. This makes for much stressful running between aircraft. Delays spread fast within the system. There can be up to five changes of crew each day. You should tell the readers that a crew that flies together all day on the same aircraft suffers less stress, less sickness. Fewer delays result, saving money.

> The key to a satisfying career in commercial aviation rests almost entirely on the airline's scheduling practices. A stable, equitable roster and the correct ratio of crews per aircraft are essential ingredients for a happy and settled life in aviation.

> The long-haul lifestyle is not healthy. We need better-quality rest and meals. This would help during flights.

> I am due to have both knee joints replaced. Is there any possible relationship between arthritis and the bodily pressure change of 10,000 feet every time I took flight over my 40-year career? Is there a cause for tinnitus (from which I have suffered since 1984) – probably as a result of my first 3,000 hours being on Yorks and Tudors with un-silenced and un-soundproofed Rolls-Royce Merlins? More medical research is probably required. Also I do not like the idea of extra-long 18 to 20-hour flights. Better to have 12-hour sectors and refuel and change crew, otherwise we might have a case of: 'I told you so ...'.

> A survey to determine the amount of depression that pilots suffer from (in silence) would be illuminating.

Still very much a taboo subject in society generally, mental health is not an easy subject to broach with pilots and their managements. The subject has been raised, however, by industry analysts like Prince (1990, p. 126): 'As if drug-taking and drinking by pilots were not intimidating enough, there is also the spectre of mental illness, which, although it must be kept in perspective, is by no means unheard of. In fact, of all the reasons for the premature retirement of pilots, mental illness is the second most common'.[3] The following testimony illustrates the kinds of pressures to which pilots may be exposed:

---

2 Isomorphic learning involves learning and applying lessons from previous incidents and accidents. See Toft and Reynolds (1997) for a full definition of isomorphic learning.

3 According to Prince (1990, p. 126): 'The mental disorder of a captain has already been confirmed as responsible for one fatal crash in modern times. The company was Japan Air

> My previous airline decided to close seven bases and get rid of my aircraft. A low-cost carrier heard about this and offered most of us positions, so ten days after the announcement I was on a 737 course. Headspinning or what? The next 12 months were the worst of my life, personally and professionally: new job; new airline (new aircraft, SOPs, rosters); living away from home for six months; we had a very traumatic house sale (one minute it was on, the next off) which virtually lost us the house we were going to move in to; my car blew up [the engine failed catastrophically] in the outside lane of the M1 motorway with my family on board; we had to cope with two deaths, one of which was very sudden. While all this was going on I was working six and three[4] and flying close to maximum hours every month. After flying 600 hours a year previously it was getting to be very hard. I started to get regular migraines at work. I was taking six to ten Solpadeine a day. I was being physically sick at home as a result of the headaches. My doctor prescribed stronger pain-killers. They had no effect. This continued for some two months until one day I had such a pain in my head that my right eye closed. I disembarked from the aircraft. My doctor signed me off work for two weeks. The relief was unbelievable! I thought I was just getting tired due to the job. So I thought nothing of it. The headaches disappeared in two to three days. While I was off sick my Class 1 medical was due. I went to see my Authorised Medical Examiner [(AME)]. I told him about my headaches. He suspended my medical for five weeks. He told me I had 'classic stress headaches'. My AME asked me what had been going on for the last two years. He was surprised that I had gone on for as long as I had done. The initial five weeks became four months, which involved a motorcycle crash and ending up in hospital for a couple of days. I felt helpless.

This pilot's story illustrates many of the characteristics of contemporary commercial aviation – its dynamism (evidenced by the fact that pilots can be made redundant and find themselves re-hired very quickly), its need to extract the maximum return from expensive human and physical capital and its capacity to engender ill-health. Having had several months off work the pilot faced a stark choice:

> The company stopped paying me ... so I felt I had to come back to work to support my family. Away from the job I felt about ninety percent 'me', but as my return to work approached the headaches came back, the stress began again and the thought of six and three, with maximum days and maximum hours, became unbearable. After nearly five months without flying they rostered me straight on to the 737-700 simulator (an aircraft I did not fly on the line) for my OPC/LPC[5] ... then a short spell of line training, then back on to the line! It all got too much. I resigned. I never wanted to walk on to a flight deck again – all this after 15 years of flying .... I did not know what I was going to do, but so long as it took the problem away I was OK. I decided to come back part-time. I am now full-time and back to full strength. Having said this, though, I still find the workload very tiring. If I could sum up what I feel about it, it is that the job of a low-cost airline pilot is very tiring, both physically and mentally. This does not leave you much in 'reserve' to

---

Lines .... The location was Tokyo Bay, in 1982 ... '.

4 That is, six days flying followed by three days off ('days off' meaning no flying and no office work).

5 The Line Proficiency Check (LPC) is a company-specific six-monthly pilot evaluation. The Operations Proficiency Check (OPC) is taken by pilots every 12 months in conjunction with a Line Proficiency Check.

> cope with the other stresses that occur in day-to-day life. You don't need much to take you over the limit. I went past my limit and tried to carry on! Incidentally when I did go back in to the simulator I was paired up with another Captain who was also coming back after three months off work due to stress.

This respondent's observation that a busy roster does not leave pilots with very much in reserve to cope with the vagaries of life is a not uncommon sentiment amongst pilots (as may be judged from respondents' and interviewees' testimony elsewhere in this book). Interestingly, while this respondent attributed his stress and ill-health to external factors, another respondent suggested that stress can be 'self-induced'. 'This is not a stressful career. However in some individuals stress is self-induced' he claimed. In connection with this observation it is noteworthy that one of the interviewees (a retired Captain) admitted that professional pride (the 'can-do' mentality) and financial consideration might influence a pilot's decision-making. By way of illustration he recalled a duty he had undertaken while flying in the United States:

> There were times, looking back on it, when I should have turned down a flight .... One time I was working in the United States where flight time limitations are something from the Dark Ages. I was supposed to be positioning from Miami to New York, then I was rostered to operate the early-morning scheduled service down to San Juan and back. This would have been fine, had it not been for flooding at New York JFK. I got a lift up from Miami to New York in one of our DC-8s. But JFK was closed because of flooding. So we ended up at Philadelphia. Then the whole thing got worse, because we were just passengers. In the end we decided to get a train from Philadelphia to New York. We arrived at about 01:00. We then got a cab to our airport hotel and got to bed at about 03:00. I was up again in about two hours' time. That was quite legal in the USA. That was the moment that I should have said: 'No way'. But there were problems with doing that. First there was no-one else to operate the service to San Juan. Secondly we were paid by the hour, so I would have said goodbye to quite a lot of money. So I went to work. Coming back in to JFK I remarked to the crew how fatigued I was. I had felt just as fatigued at other times but that occurrence was entirely my own fault. I should have said 'No'. But, you know, you do carry on. There was another occasion when I should have declined a duty. We were operating Boeing 767 flights out of Chicago to Rome. I discovered that not only had I been ill all day, but the First Officer had also been ill. We had both contracted some sort of stomach bug. When we turned up at the airport we both felt absolutely awful, even thought we had rested in bed all day. The nearest alternative crew was about 4,000 miles away, and we were operating a scheduled service. I would say that on that occasion I was foolish in not saying: 'I'm sorry, I cannot do it'. Anyway, we did the duty. Surprisingly by the time we reached Rome we both felt absolutely fine. But I should have said 'No'.

The retiree offered a theory as to why flight crew may place themselves, or allow themselves to be placed, in a potentially risky situation:

> What happens is that you manoeuvre yourself – or are manoeuvred – in to a position where by saying 'No' you let down several hundred people, if not several thousand. And of course part of the problem with feeling ill is that you are not making good decisions.

> For example, a Number One [a senior member of the cabin crew] fell over on one occasion and hit her head. She said 'I'm fine'. I said: 'You are not going to carry on. You are going to go home'. Later she went into shock.

The retiree was asked whether he believed flight crew had a 'can-do' mentality?:

> Yes. The biggest element of wanting to carry on is that you are frightened of screwing up [being seen to give up on a task]. The astronauts came out with that observation [see Wolfe's *The Right Stuff*]. Getting killed is tough, you know, but getting caught screwing up is something you want to avoid. You really want to push as hard as you can to get the job done. That can be a real problem. Of course it is just what you need in the military. You also need a certain amount of it in civil aviation as well, otherwise you would never get anywhere. If you really needed to you could always find a good reason for not going, but you gear yourself to getting over these problems and carrying on.

He went on to claim that taking a 'managed' sleep on the flight deck might offer a way of countering the effects of fatigue:

> Under the British rules they accept that you can cat-nap. Under the American rules you are not supposed to. I have actually fallen asleep with the FAA chap sitting behind me ... the problem is that sleeping is a physiological need ... I do not think that you can say to people that you will not take a nap, any more than you can say you will not go to the toilet ... if you do not do it in a controlled, voluntary way then it is going to happen in an involuntary way. So I have always believed in taking a nap in a managed way. Sometimes you will find that your colleague has had a hell of a day, the children have been ill, etc., and this chap has turned up at work having had no rest. Absolutely no problem. You can cope with that because you know about it. You can say: 'OK. Get your head down. I will wake you up'. The same can happen the other way around. It is all sensible management, and you can establish a relationship and you can sort anything out like that. It is when people do not tell you that you have problems .... [When I did fall asleep] it was not always at night. I could drop off inadvertently once or twice in the daytime. It's warm and sunny on the flight deck. You are sitting there relaxed, and you suddenly realise that you have dropped off to sleep. It can happen inadvertently. Food can play a part in the process. But then you have got to eat. Some nights you are fine and you stay awake. Other nights you drop off. Some flights pass very quickly. I cannot explain why sometimes you feel fatigued, and at other times you feel absolutely fine. In theory it is to do with the amount of rest you have had previous to the flight ... but I was always good at resting prior to my duty.

This interviewee also mentioned that he took sleeping tablets to help him rest before a duty: 'I was always prepared to take something to help me sleep. I started off taking Temazepan ... I would go to bed in the afternoon and the sleeping pill would "push me over the edge". I would always try to get some sleep before a night flight. But even then it did not always work. However, it is important to try to get some sleep. It is important to consider how the inquiry lawyers would look at what you did, should there be an investigation of some sort'. It might also be suggested that it is important to consider the latency of the effects of individual sleeping drafts (could they, for example, last so long as to impair judgement on the flight deck?) and their long-term health impacts and efficacy. Another interviewee stated that a pilot's decision-making could be impacted by colleagues' performance and/or management pressure:

> We were on a late duty ... four sectors with fog across the Midlands. We were trying to get into EMA. We went in to the hold for a while, then we diverted to Birmingham. It was getting late by now. We went to the Crew Room and spoke to Operations in Aberdeen. We informed them that we were leaving our aircraft at Birmingham. We were then put under immense pressure to get the aircraft back to EMA [their home base]. Additionally we were an all-female crew. It was like all the other crews had got in to EMA and the girls had not. We resisted and resisted ... I was completely and utterly shattered ... because it is very stressful hanging around in fog on top of a big day and I was saying: 'I am not going back'. Eventually my Captain caved-in, and then I gave in. But I told her that I would not handle [be the pilot flying (PF)] on the way back, even though it was my sector. Psychologically it was really hard because we had left the aircraft, collected all our things, gone to the Crew Room. We were thinking that we would get in there, order a taxi and get home. As it was we had to return to the aircraft after spending about sixty minutes in the Crew Room.

Of course pilots are not the only professional group to find themselves being worked hard by management. National Health Service junior doctors, for example, may find themselves working extended duties for many years. Regarding the length of the working week in the UK, Summerfield and Gill (2005, p. 54) note: 'The length of the working week varies by occupation ... 40 per cent of managers and senior officials worked over 45 hours a week – the occupational group most likely to work these long hours .... Government policy over recent years has stressed the importance of maintaining a healthy work–life balance'. Like the Government, the General Secretary of the Trades Union Congress has also drawn attention to the need to maintain a satisfactory work-life balance. He has also sought to highlight the wider social impacts of long working hours: 'Increasingly, people are balancing the demands of caring not only for children but also dependent adults, and these workers need support and protection. But long hours working is the biggest demon facing UK workers. Many fathers find themselves spending extra hours at work when they would really rather be at home, which in turn forces their partners to reduce their hours and pay to run the home and look after children. A better work–life balance where men and women could spend more time with their families and be less stressed at work would be in everyone's interest' (Barber cited in Osborne 2005a). While there may indeed be a greater focus in society at large on maintaining a healthy work–life balance, there was a clear perception amongst both respondents and interviewees that they were being pushed to the limit of their endurance. The following testimony from one of the two female interviewees is informative:

> I was doing night-mail flying up in Scotland. You started your duty at 22:00 and ended up in Kinloss at 03:00. Then you flew back to Aberdeen at 06:00. So you did the work-arousal bit, then between 03:00 and 06:00 your body-clock shuts down ... nothing to do ...

> brain shuts down ... then you flew back to Aberdeen. On this occasion we had low cloud so we could not see the runway. We had to do a go-around. Me and the Captain did not work well together as a team. We concluded that we should have decided to go around before we did. But we were both so tired we just wanted to get on to the ground. That is the only occasion I can think of when both of us made mistakes because we were both so tired. I have flown quite a lot when I am exhausted, but generally you and your colleague are not at the same point of exhaustion ... that is the only time I can think of where both of us were not one hundred percent .... [When you are fatigued] your concentration goes. I find it an effort to think about what I am doing ... it takes me longer to read the plates [check-lists] and I have to really search for information. Setting things is a conscious thing rather than using motor skills. I have to think: 'I am going to do that by doing this'. It is quite noticeable. I can tell when I am too tired, which I hope is a good thing because you can recognise it yourself. It is not a subconscious slipping into being rubbish. Also I get thirsty. Drinking water seems to keep me more alert.

The interviewee was asked why she thought she and her colleague were not working together particularly well prior to the night-mail incident:

> Because he was exhausted as well. We knew each other quite well, and I think we were both shutting down and not telling each other we were shutting down. I was fairly new to commercial flying then, and I think both of us shared the same mind-set of: 'I am so tired I just want to go home'. We only did that particular duty every two weeks and the rest of the time we flew day-time, so the night-flying was a bit difficult because we were not used to it ... we did not know what to expect.

Asked what she and her colleague would do between 03:00 and 06:00 during the night-flying duty, the interviewee replied:

> We were taken to a public house where the landlord rented out his lounge room. We sat there and could drink tea and coffee and watch TV. You could sleep on the sofa if you wished, or use a little bedroom at the end of the hall. But you only had three hours during which time you had to get a taxi to the pub and back again. I found that if I went to sleep I felt worse when I woke up. I need more than two hours sleep. I either need ten minutes or eight hours. I cannot do a two-hour sleep. Generally most stayed awake and watched TV.

This testimony illustrates two of the chief characteristics of commercial flying today. First, that it can be fatiguing – to the point where a pilot's ability to make high-quality decisions is impaired, and the quality of her/his home life is reduced. Secondly that commercial flying can be decidedly unglamorous. With regard to mail operations of the type described above, Kohn's (2005, p. 28) observation is germane: 'Some of the less sociable operations that carry newspapers by night are less appealing because of the late in the day flying, night after night'. One is also reminded of Gandt's (1999, p. 4) description of the USAF Hercules pilot's reverie: 'Chuck Kraft ... had spent the past four years eating box lunches and drinking bug juice and flying Air Force C-130 transports around the world. During his missions a scene had begun to repeat itself. At each layover – Madrid, Paris, Bangkok, Hong Kong – he would head directly for the hotel where the airline crews stayed. At the swimming pool he would observe the same curious sight: *Pan Am Stewardesses* – Swedish, German, French, American – gorgeous in their bikinis, frolicking around the pool. Frolicking with them would be smiling, suntanned, obviously overpaid *Pan Am pilots*. One day in Bangkok, standing there in his evil-smelling, baggy flight suit ... Kraft suddenly fathomed an immense truth: *They're doing the same goddamn job I am*'.

*Airline managements: in or out of touch?*

There was a feeling that managements were failing to engage with flight crew, and consequently that they knew too little about the 'front-line' and impact of their decisions. As one respondent wrote: 'Pilots' opinions are not wanted. Air traffic controllers have their air space changed without consultation. Managers stay behind desks, cut off from reality. They are becoming "super clerks" (the General in the Château syndrome)'. Another said: 'I am paid £100,000 per annum but I definitely do not feel valued as an employee. My boss ignores me so there is no communication. I feel little loyalty to my employer'. Some interviewees gave practical examples of how management decisions taken (in their opinion) without adequate consultation could make the job more difficult. The following comments were made with respect to standards of ground-handling:

> [M]y airline was cutting costs on ground-handling big-time, and it showed ... the classic situation was coming on to stand with nobody to meet you in terms of getting carry-on baggage out of the hold. Passengers were not allowed to leave the aircraft until the hold had been opened and the carry-on luggage brought to the bottom of the steps .... One day at EMA we came on to stand and there was no-one to meet us ... so we waited ... and eventually the Captain and I did it ourselves .... I levered myself into the hold ... in my uniform ... and we emptied the hold of the outsize carry-on baggage ourselves ... we were more than happy to do it just to get the passengers on their way [an example of pilots' 'can-do' attitude].

A second interviewee highlighted the impact of cost-cutting on ground-handling:

> There is more time-pressure. Cutting costs means fewer staff. In my last short-haul company that might mean Dispatchers doubling-up on shifts. Perhaps working late-to-early shifts with only a six-hour break in-between. There might be fewer Dispatchers dealing with more flights. Also there is no accredited training for Dispatchers in the UK. [Things could be improved] through accredited training. The problem is that it is a low-status job. Wherever I have been there is a high turnover of staff. The industry does not treat people well, especially in regards to shift-work. The industry could address the turnover issue by paying them more, and working them less.

A third interviewee compared standards of ground-handling in the UK with those she had experienced abroad:

> Ground-handling is appalling here [in the UK] in comparison to overseas. In my experience there is a complete lack of interest in the job. The people they employ are not given enough training. They are not given enough bodies to do the job properly, so it's probably not all their fault ... but there is an unwillingness to help ... it is oceans apart from what you experience overseas ... there are a lot of people in the job who do not seem to care .... There was this one chap who would sit at the rear of the aircraft and chat-up the cabin crew ... and not bother to come and speak to us [the flight crew] ... he did not bring the weather to us ... did not give us any information. On one occasion we missed a slot because he did not tell us we had one ... he was rubbish at the fuel and getting the passengers out on time.

There was clearly a feeling that management decisions, often taken with a minimum of consultation, were having a detrimental impact on ground-handling and other operational areas. One respondent was moved to exclaim: 'The industry must listen to pilots!' Feedback mechanisms are now a common feature of industrial management. Even the automotive industry, for so long plagued by worker-management mistrust and antagonism, has adopted a more enlightened approach to management. Bartol and Martin (1998, p. 463) use the example of Detroit Diesel to demonstrate the potential of what they term 'New Management':

> In the new culture, *managers seek information from the workforce*. One technique used at Detroit Diesel that often helps managers from becoming isolated is 'management by wandering around (MBWA)', a practice whereby managers frequently tour areas for which they are responsible, talk to various employees, and encourage upward communication .... Does this system work? Over a 4-year period, Detroit Diesel increased its share of the market from 3.2 to 28 percent because of the improved quality of the engines produced by its work force – which now speaks out. (Emphasis added)

While feedback loops are not unknown in commercial aviation – for example, the UK's anonymous and neutral Confidential Human Factors Incident Reporting Programme (CHIRP) – there is a significant difference between maintaining a passive report-back system like CHIRP and leaving your office to spend time on the shop floor (i.e. the flight deck, aircraft cabin, ramp, crew room, training centre, simulator hangar, engineering shop, call centre, check-in desks, baggage undercroft, etc.) asking questions and encouraging comment.

### *Other worries – from the 'transformative' impact of LCCs to deskilling*

Besides being concerned about commercial pressure and managers' remoteness, interviewees and respondents were also worried about: the impact of the LCC sector on the industry at large; the prevalence of sex-discrimination, and its impact on the recruitment and career-paths of female pilots; the impact of advanced cockpit automation on pilots' traditional hand-flying and problem-solving skills; the apparent failure of many employers to provide for the nutritional needs of flight crew; declining experience levels amongst line pilots; the failure by some elements of the aviation system to use only the English language to direct operations; and the inception of a more 'aggressive' style of airline management.

*Transformative LCCs*

While some respondents lauded LCCs, others had serious reservations about the low-cost phenomenon. Some doubted the robustness of the LCC model's financial component. Typical comments were: 'I think some LCCs will be taken over or go out of business. I also think many passengers will see the myth of "low-cost" and return to the full-service airlines (I hope)'; 'More LCCs will start then fail'; 'I foresee public disenchantment with the low-cost model, and a return to a more stable environment with a solid economic base [this pilot flew for a low-cost airline]'; 'Lots of LCCs will go bankrupt or be bought out'; '[There is] insecurity caused by low-cost airlines'; 'The LCCs will compete with, and drive each other into the ground ... '. The respondent who made this last comment warned that a major accident was likely '... given the training and recruitment trends, coupled with fatigue issues'. Another respondent noted: 'We must recognise that low-cost, high-intensity flying with very short turnarounds is much more fatigue-producing than conventional aviation work patterns'. (The stress and fatigue-inducing potential of busy rosters was a major concern amongst respondents.)

Other respondents cast the LCC sector in a more positive light. Several constructed LCCs as instruments of change. One said: 'I foresee growth of the low-cost market, with a corresponding change towards niche marketing for the charter airlines. The majors will have to adjust their own business models in order to compete effectively with the LCCs, or move away from direct competition and concentrate on long-haul'. Another foresaw: '[C]onsolidation of the majors, big layoffs, the rise of low-cost carriers and the decline in general of amenities'. One respondent was sufficiently enamoured to say: 'I would like to be working for a low cost carrier'.

*Sexism*

There was a feeling that the industry was still struggling to encourage and integrate female pilots. Some claimed to have experienced overt discrimination on grounds of gender. Each of the two female interviewees claimed to have suffered discrimination from an instructor:

> [I have not experienced a negative reaction] from a flight-deck colleague. I have had a negative reaction from a simulator instructor, though. I had just finished my initial training and was progressing to line-oriented flight training [LOFT] conversion in 1997 .... He was an ex-BA Training Captain, knocking-on 64 years of age, he'd worked in Singapore for five years then retired from there .... He was 'old school' .... He actually said to me: 'Women should not be flying'. I had difficulties with him in the simulator to the extent that he had to be removed as my instructor. He was picking on me quite substantially.

Asked if she had ever challenged his attitude, she replied: 'I was in a difficult position because I was an airline-sponsored cadet. I was not really in a position where I could fight my corner'. Asked if she had had similar experiences on the line, she replied: 'No, because most of the people I have worked with have been pretty much "You're a pilot", which is good. I'm sure that some of them have cut me a bit more slack, possibly'. The other female interviewee said:

> [I have not experienced sexism] in an aircraft. But I have had a reaction in the simulator from one Training Captain. When I flew with him on the line he was fine. We would have a good day together. In the simulator I believe he was unfair to me. I think that was down to my gender. No matter what I did it was wrong .... I would be told off for not being forceful enough ... then I would be told off for being forceful ... then I was 'too assertive', 'overpowering'. I thought 'I cannot do anything right, so I'll just try and get through this'. I would not learn anything as a result. This to me was a sin, because you do not get much time in the simulator, and you desperately want to go in there not just to pass your test and get your licence renewed, but to learn; be trained again; come out feeling confident that if you have an engine failure tomorrow you will be on top of the situation .... It was always a great shame if I got him in the simulator because I used to come out feeling really low.[6] [This Training Captain] was probably in his mid-fifties. He had thousands of hours. He had had a light aircraft crash.

Others felt that discrimination could take more subtle forms. One LCC-employed respondent, who had encountered only 'limited animosity' from male colleagues, claimed that the pilot-lifestyle tended to work against relationships and women's interests generally:

> As a female pilot in the aviation industry I am very lucky to have encountered limited animosity or prejudice. However the job itself poses its own problems: women who work often go on to marry and have children, but many find it increasingly difficult to do both. I am neither married nor a parent, but the chance of doing either is made that much harder because of my career. My current boyfriend is also a pilot but works for a different airline. He is based miles away from me and consequently we never see each other. Even if I ended up living with him we would still have to contend with the fact that with different roster patterns it would be hard to plan our lives or get much time together.

The difficulties experienced by pilots with regard to maintaining a relationship have been documented elsewhere, as with this First Officer's testimony published on the British Airways (2005) bulletin board *Explore Our Working World*: 'We could be doing four flights a day to destinations all over the UK and Europe, and by the end of it you are exhausted .... I'm going out with someone from the airline. At times our scheduling patterns mean we only get one day off a month together ... '.

Regarding the possibility of having a family while holding down a full-time flying job, the LCC-employed respondent continued:

---

6 Despite this outcome the interviewee said she had never made a formal complaint against the Training Captain.

When the idea of children is thrown into the equation it gets even more complicated. Very few airlines offer part-time work for their pilots, probably because most are men. Few child-care facilities open at 04:30 and close at 02:30, for example. One either needs a full-time nanny, extremely helpful grandparents or a partner who does not work. Add to this the possibility that the female in question might be approaching a career goal at the time of wanting a family. Perhaps taking two years off instead of going for one's command is not wise. These are problems that affect just women, never mind men. I am sure that the aviation industry has one of the highest divorce rates of most professions, and quite possibly illegitimate children. The long hours away from home in hotels around the world have their consequences. Short-haul flying cannot cure this as most LCC pilots spend more time at work than at home. There is tremendous strain from pure flying hours worked with LCCs. Indeed many of my single male friends cannot even start a relationship as the only people they have time to see are cabin crew, and even then there is barely time on a turnaround to say 'hello'. As much as I love my job I did not think it would be at the expense of a social life. Admittedly you may get different feedback from pilots with other airlines. This is just my limited experience. I do not want to say that this working environment is any harder than any other profession. It is just different. In my view many people do not realise how different it is until they are directly involved. [This 27 year old respondent had ten GCSEs, four A-Levels and an upper second class Honours degree. She flew for a UK-registered LCC.]

Writing in *The Log*, Ball makes the following claim: 'I tried to ascertain if sexism or racism is blighting the working lives of our minority pilots .... From a small and unscientifically, but randomly, chosen cross section, no-one could relate an incident where they felt that they had been discriminated against in their aviation career' (Ball 2005, p. 6). Clearly Ball has not considered how the pilot-lifestyle and general lack of social support for working mothers might amount to *de facto* discrimination. As to the question of sexist attitudes amongst male pilots the following testimony is illuminating: 'I'm not comfortable with female pilots, but that's probably because I was born in 1940. I know they are slowly taking over the industry .... I suspect that in 20 years time we'll have more than fifty percent female pilots ... I've heard it said, when a female voice comes over the radio: "There's another empty kitchen"' (this interviewee had flown with a female pilot). By way of contrast another interviewee claimed he had 'hardly ever' encountered sexist attitudes on the flight deck. Asked to explain this he stated:

Well, it's the nature of the task, isn't it? It is all about safety ... being sexist would impede communication. If you display a prejudice against somebody it is going to be difficult to work with them. You are stuck together in a chicken coop that is travelling at five hundred miles per hour. If you do have prejudices you should leave them outside the cockpit. The same thing applies to the cabin crew. If you are going to create a barrier to communication they are not going to tell you that the rear of the aircraft is ablaze.

Another interviewee said that he considered female pilots to be 'just another member of the crew'. Asked if he had ever behaved differently towards a colleague on grounds of gender he said:

> Yes. Quite simply, if I happened to accidentally contact the female on her chest or behind [buttocks], for example, I would apologise immediately. Otherwise they are just another member of the crew. I have flown with some very good female pilots.

It can be seen that interviewees and respondents had mixed views on the prevalence of sexism on the flight deck and in the industry generally. More research is required before an accurate assessment can be made. It is clear, however, that deficiencies in social provision conspire to impede women's progress. London-based pilots with early duties at Gatwick Airport, for example, are obliged to rise as early as 02:30. There are no crèche facilities available during the very early morning. In the UK the lack of childcare provision has long impeded women's progress in the jobs market. Hudson and Williams (1989, p. 114) note: 'One of the major constraints on the lives of all mothers in the UK has been the consistent failure to develop child-care facilities'. For most of its history the aviation industry has been dominated by men. The industry needs to consider whether, in terms of giving *practical* help to women employees, it is keeping pace with other sectors. Bartol and Martin's (1998, p. 342) summary of developments in the United States, and the impact of those developments on labour relations is illuminating:

> Given the rising proportion of women in the workplace, the growing number of dual-career couples, and the frequency of single parenting, family issues as they affect workers are increasing in importance. To help employees more effectively handle the often conflicting responsibilities of work and family, several major companies, such as Du Pont, now have family issues specialists on their human resource staffs .... This type of approach is likely to promote positive working relationships with employees and facilitate the management function of leading.

### *Automation*

As for concerns about advanced automation, some respondents feared they would lose their hand-flying skills. There was a feeling that, with the advent of the highly automated glass cockpit, pilots were being transformed into systems managers. As one respondent put it: 'As automation increases, airmanship declines'. Of course there are those who cast advanced automation in a positive light. This testimony, taken from the joint Royal Aeronautical Society and Guild of Air Pilots and Navigators publication *So You Want To Be A Pilot 2006*, is typical: 'The "job" of a pilot has changed over the last 10 years as aircraft have become more automated and the skies busier. As well as being able to demonstrate good airmanship, pilots need to be able to take commercial decisions and think outside the flight deck. All in all, the job of a pilot has evolved and in my opinion this has made it more satisfying as a result' (Basnett cited in Kohn 2005, p. 43).

Modern glass cockpit aircraft require not only that pilots develop systems management skills but also that they are comfortable working on a two-person flight deck. One interviewee regretted the loss of the engineer. As he explained: 'When you set your speeds for take-off it is not easy to see from one side of the flight-deck to the other the speeds the other fellow has set ... the flight engineer can check this ... and it is surprising how often he identifies that you do not have the same speeds ... it takes a lot of strain off the pilots. In fact it is excellent to fly with a flight engineer who is putting his back into it as well. Very helpful. Definitely an asset'. Paradoxically this interviewee also believed that having a third person on the flight deck could, on occasion, create tension. As he put it: 'You could have a situation where two of you are working against the other one .... I have seen a bit of that since I have not been in command of the aeroplane [as the interviewee was over 60 years of age, he had been relegated to the right-hand seat] ... you can get two of you talking about the other one when he is out of the office [colloquialism for flight deck] which is not satisfactory'.

*Food*

There was some comment about food and diet. Knowing that their career depended upon passing a periodic medical, most pilots took some form of exercise. Some were also aware of the need to eat healthily and regularly. Several pilots had concerns about the availability and quality of food, both because it impacted their long-term health (and therefore their career prospects) and because it impacted performance. Interviewees were asked why skipping meals was a problem. One said: 'Because you get light-headed and you feel uncomfortable ... '. Another said ' ... my concentration goes. I find it hard work ... '. A third said: 'Access to food and fluids is essential for performance'. An interviewee who flew for a freighting company said:

> Until very recently, apart from one of our jobs, there was no provision for food at all. There are various machines located around the various rest-rooms [dispensing snacks]. When we go through Paris, the French Postal Service provide three food trays for the crew ....

Another interviewee bemoaned the quality of food carried by some crew room vending machines: '[In the best] there is fresh fruit salad and proper, decent sandwiches .... Worst ... normal vending-machine stuff ... just not good'. On-board meals were also criticised. 'Usually the problem was having too much food ... too much fat and grease and not enough fruit' said one interviewee. Another said that her airline's cost-cutting had inhibited access to food:

> At first the airline would put crew meals on for us, then that stopped. Then instead they would load some extra passenger meals. So you could almost guarantee you would get something to eat. But then they stopped doing that. They cut catering to the knuckle. Then there was only a slim chance you would get anything. You would get something only if a passenger declined food. There was no way you could leave the aircraft during a turnaround and nip to the shop. So I always took sandwiches because I could not perform unless I had food inside me. However a lot of my colleagues would hope that they would get something, and if they did not they might sneak a packet of biscuits ... but I could not do a day's work without something inside me.

Aware of problems to do with the availability, quality and timing of crew snacks or meals, pilots developed 'work-arounds' such as eating a large meal before flying. As one freighting pilot explained:

> On-board we have a kind of tool-box full of biscuits, fizzy drinks and packets of soup, so I do not tend to open that unless I am in need of a 'sugar-rush'. It's a load of rubbish. If you do not get off the aircraft this is the only access to food that you have. It is not always possible to take food with you because you go away on tours. It is not as if you go home every night. You can go and find a supermarket but sometimes I do not wake up in time ... so you have to rely on having something in the hotel .... Because it's night freight your body clock is all gone anyway .... I tend to eat a big meal before I go to work ... generally two hours before I start flying. If we have a 23:00 report I would eat at about 21:00. There is an opportunity down-route during longer turnarounds to eat food available from vending machines .... In Cologne there is a German sausage wagon that's parked up .... You're not keen to join [my airline], are you? You'd be surprised how desperate you are for that German sausage, though!

The work-around of eating a meal before your duty carried certain risks, however. One interviewee recalled how eating a hotel meal led to a potentially dangerous in-flight situation:

> I once had a three-day stopover in Accra in Ghana. I ate in the hotel restaurant at 18:00 because I trusted the hotel food. I had omelette and chips. I then reported for duty at about 21:00 for a six-hour sector back to the UK. I felt somewhat queasy and light-headed during take-off and then 25 minutes later at top-of-climb I was violently sick. I then became dehydrated. I drank water. My Captain was not going to take us back to Ghana. It was not the best place to be ill. Equally I was not much use for the rest of the flight, so it was almost single-pilot operation. I sat there feeling dreadful. I was off sick for a week with gastro-enteritis.

For another interviewee the problem was missing his usual pre-duty meal: 'I usually ate beforehand ... I was always aware of the need to maintain my blood-sugar level ... I had an interest in aviation medicine. On one occasion I had an argument with the Number One. Looking back on it now I think that I was being unreasonable, probably because I had not eaten. On one occasion I refused to do a flight because there was no food available. I was working for a large Italian international carrier on a contract. The airline's idea of food was some ham and a bread roll. We were scheduled to fly to Chicago. I said to the airline: "If you want us to do this flight we must have food". I managed to get meals put on board the aircraft for the crew'.

Despite understanding the link between regular eating and performance, and regular eating and long-term health, some pilots confessed to wilfully missing meals: 'You quite often go for long periods without food because you are too busy, or you forget, or you do not realise you are hungry because there is too much adrenaline going round in your system'. One interviewee admitted to having gone a full day without a substantive meal: 'This happened during training. I was with this one Captain for the whole week. We made a joke of the fact that we were only eating once every 24 hours. At home I have at least three meals a day, and I thought I would be starving-hungry, but I wasn't. Very odd. I have not gone longer than that, yet'.

*Pilots' experience levels*

One respondent claimed that experience levels on the flight deck were being eroded. As he put it: 'Lower experience levels were becoming a problem with an increasing failure rate on command upgrades, mainly due to the unavoidable fact that the FO's handling skills were suffering with the advent of ultra-long-haul routes'. Another respondent expressed the opinion that: 'Airlines should employ pilots with 2,000 to 3,000 hours experience ahead of low-hours college-qualified types'. (This respondent had bemoaned 'Pilots joining airlines with no general aviation experience, i.e. not serving an apprenticeship and expecting command with only limited experience'.)

Certainly the rapid expansion of airlines operating in the Pacific Rim has placed a premium on experienced pilots. The United Kingdom Flight Safety Committee's 2005 seminar 'Aviation Safety: Looking Ahead 20 Years' addressed the issue of reduced experience levels on the flight deck. The Federal Aviation Administration's Chief Scientific and Technical Adviser for Flight Deck Human Factors noted two trends. First, the operationalisation of pilots with lower experience levels. Secondly, a greater reliance by airlines and pilots on prescribed procedures (SOPs). 'I'm not saying that's wrong', said the FAA staffer, 'but there needs to be a trade-off to ensure pilots can handle unexpected situations. There are various examples where flight crew have saved the day, such as the Transair A330, Iraqi A300 and Malaysian 777 incidents' (Abbott cited in Sheppard 2005, p. 8).

*Whither the lingua franca?*

One respondent expressed his dismay at the use of languages other than English during flying operations: 'With the present volumes of air traffic, I am concerned that there is still not a common language between air traffic controllers and pilots at major international airports. For example, the French always speak French to each other. I had a close airmiss over the centre of Paris – fortunately my Traffic Alert and Collision Avoidance System saved me. It is permissible to fly with this system unserviceable. The other aircraft involved was French. I thus had no idea where he was or what he was doing. If we had collided over the centre of Paris thousands of people could have been killed. Although I cited language as a mitigating factor in this incident, the French authorities would not recognise this in their final report. If the French have to be so intransigent that their language is worth more than thousands of [French] lives, then there should be a law either making French the international language or ensuring that all pilots operating in France should be able to speak French. The language issue has already caused a runway collision at Paris Charles DeGaulle, resulting in the death of a pilot'.

### *Management style*

Some respondents alleged that airlines were adopting an increasingly aggressive style of management. One said: 'It's a great job, but there are rough seas ahead. It is hard to compete with Ryanair and their management-by-fear. Any Ryanair pilot that questions management will be transferred to a base far away with two weeks' notice. There seem to be new LCCs every day that are willing to fly for less. Time will tell what the skies will look like'. Another urged that '... bully tactics from employers should be forbidden by law'. A third alleged that managements could be tempted to use pilots' emotional attachment to flying to manipulate them: 'One of the most interesting factors that comes into play in the flying industry is that – unlike many other professions – pilots come into flying generally because they have a great love for it, prior to any thoughts of income. This affects all matters of contract negotiations and employee management. It is more than just a job, and while because of this it is the most rewarding career I could imagine, management realises this and uses it to their advantage against pilots who are just starting out'. Concerning this last point it could, of course, be argued that all industries seek to exploit their employees' enthusiasm (until, that is, it dries up). Consider, for example, how vocationalists such as doctors, nurses, teachers and police officers have been manipulated in the past.

## Summary

There was a distinct feeling amongst respondents and interviewees that efforts to get the maximum return possible on human and non-human capital might be impacting safety margins. Some believed that certain airlines had a distorted view of FTLs, alleging that these airlines saw them more as targets than 'limits of last resort'. Several respondents highlighted the work-rate in the low-cost sector, expressing the view that it could become the industry norm. While it could be argued that pushing the envelope is a more-or-less inevitable consequence of free market competition, it should be remembered that aviation mishaps can have catastrophic outcomes. Hundreds can be killed. Because of the 'dread' character of aviation risk, every accident deals the industry a severe blow. Thanks to hyperbolic media coverage the public is often left with a distorted impression of the industry that further intensifies feelings of dread.

Airline managements were roundly criticised. They were perceived by many as remote and uncaring. While this refrain is not uncommon in industry generally, it is a surprising development in an industry whose front-line workers (pilots) are such capable problem-solvers. Given that they take so long to train, cost so much to employ, spend so much time at the 'coal face' and are able to solve difficult problems under pressure, why not tap (that is, exploit) pilots' problem-solving skills, familiarity with the 'lived reality' of flying and accumulated wisdom?

Airline managements were also perceived as exploitative and interested only in the 'bottom line'. Again this refrain is quite common amongst employees and their representatives. Consider, for example, this assessment of the contemporary business ethic from the General Secretary of the UK's Trades Union Congress (TUC): '[Today organisations] exist to deliver value for shareholders and everything else takes second place. As long as "value" is being squeezed from a business then investors will not pay too much attention to how the organisation is managed .... [C]orporate managers need to keep the share price high. And how do you do that? The simple answer is that you downsize, outsource or look for a merger or acquisition' (Barber 2002). Of course, corporate managers might retort that delivering a healthy return is the only way to safeguard employment.

Respondents and interviewees reflected at length on the impact of their working life on their home life. Many believed the former had impacted the latter. Some attributed failed marriages to work, either because it drained them to the point where they simply let the marriage disintegrate, or because it put temptation in their way (time spent away from home with attractive companions). One respondent said: 'Lots of people comment on the work and lives of commercial pilots and some of them make the rules up by which we live and work. These are largely from men and women in grey suits in cosy offices with a 9–5 existence and weekends off .... I would very much like these suits to hang around on jump seats and work our rosters with us a bit more, and see the effect they have on our lives at work and at home'. Similar sentiments have been expressed elsewhere, as with this statement from a Boeing 747 First Officer ('Gary') taken from the British Airways (2005) bulletin board *Explore Our Working World*: 'There's nothing glamorous about getting up at 3am ... a lot of what we do is pretty hard work. Because of the working patterns, I'm often on a completely different time zone to my friends .... Of course, my wife misses me when I'm away, but when I'm home she's had enough of me after about three days and is relieved when I go off to work again!'

Finally, a comment about pilot-culture. It was clear from their statements that respondents and interviewees got a great deal of personal satisfaction from their work. For these pilots, flying had a visceral, transcendent quality. Again this sentiment has been expressed by others. As one British Airways Boeing 757 Captain ('John') wrote on his company's website: 'As a pilot you feel part of the plane .... There's nothing to beat climbing up through the cloud first thing in the morning on a grey day with dawn breaking and seeing all those wonderful colours. It is inspirational .... I love what I do and even if I won the lottery tomorrow, I'd go on flying 757s' (British Airways 2005). Pilots' emotional engagement with their work is part of the pilot-culture.

# Chapter 7

# Pilots' Aspirations and Future Plans

The most common aspirations were: to get a promotion (to either Captain or Training Captain); to get involved in the training of more junior pilots (as one respondent explained, so he could 'put something back in'); to change to a different type of flying (some wished to fly long-haul, others short-haul); to move to a part-time contract (sometimes because the respondent wanted a lighter workload); to get involved in airline management; to fly large, modern, 'cutting edge' aircraft (the Airbus A380 was mentioned); to fly small aircraft (there may have been a perception that small aircraft make more demands on a pilot's 'stick-and-rudder' flying skills – skills that pilots seemed proud to have acquired); to secure more relaxed rosters and more time off; to get through the next simulator session, recurrent aeromedical examination and check-ride (for some respondents such checks seemed to generate a degree of psychological pressure).

It was clear that pilots wanted different things from their employer and career. To some degree what a pilot wanted reflected her/his personal circumstances and time spent in the profession. While some were attracted to long-haul operations others wanted the relative stability of short-haul flying where the duty ended where it began – back at base. In situations where pilot morale is low, managements might consider trying to match rosters and career paths to pilots' (varied) aspirations.[1] Summerfield and Gill (2005, p. 58) observe: 'Most people who want to move jobs do so because they are dissatisfied with aspects of their current job [in the UK in 2004, 26% of those looking to move jobs cited unsatisfactory pay as the reason, 8% wanted shorter hours]. In spring 2004, 6 per cent of both male and female full-time employees in the United Kingdom were looking for a new job'.

Of course, pilots' ability to satisfy their ambitions is in large part a function of the state of the industry. Industry recessions, such as that which occurred after the terrorist attacks of September 11th 2001, stymie opportunity. Pilots are laid off in large numbers. Airlines go to the wall (unless, of course, they are US majors or European flag carriers, in which case they either go into bankruptcy protection or are bailed out by their governments). Thankfully, however, there are opportunities even in the gloomiest of recessions. While US majors have been gliding in and out of Chapter 11 bankruptcy protection, new opportunities have been emerging in countries like India and China – by common agreement two of the emerging powers

---

1 Given management's need to maximise the return on physical and human capital there are limits to how far flight crew preferences can be accommodated. An airline board has several key objectives, one being the maximisation of shareholder value.

of the New Millennium. Both countries are developing their aviation infrastructure to encourage air travel. India's air travel industry is growing by leaps and bounds. Bhargava (2005, p. 53) puts it in perspective: '[I]n 2004, India posted a high GDP ... annual growth rate of around 7%. Companies the world over were outsourcing business processes (such as call-centres) to India and exports were healthy for software and services, all of which resulted in the young middle class suddenly having considerably more spare income. These dramatic changes brought about a surge in air travel, with numbers increasing by 22% and 26% per annum [of course this "surge" would have emanated from a relatively low base figure]. Almost all aviation experts in India now forecast that the high rate of increase in air travel is likely to continue for a long time. The number of Indian passengers is expected to grow to 50 million in five years as the new low-cost carriers stimulate demand. The scope for growth in the air travel business in India is significant and is ripe for exploitation by entrepreneurs'. According to Tarry (2005, p. 72), Asia '... is the area where there is both momentum and potential'. Clearly there are opportunities for both pilots and investors in such buoyant markets. Having said this, it is never easy to relocate to another country to find work, especially if one has put down roots at home. As a depersonalised economic theory, global labour mobility looks attractive. As a lived experience it looks rather unattractive. Migrant working can be a demoralising experience.

Generally, the younger the respondent the less s/he had thought about retirement. Of those who had pondered retirement, some had decided to retire as early as possible while others expressed a wish to work for as long as possible.[2] There was some resentment that pilots had to retire at a relatively young age. One 57 year old respondent exclaimed: 'I aim not to retire ... I'll be forced to retire soon as I fly single-pilot public transport. Am attempting to change the legislation!' There was also a degree of fear. One respondent said: 'I'm not sure if I am looking forward to it. We [presumably pilots] seem to pass away very quickly after retirement. Perhaps we need the stimulation'.

Retirement plans included: farming; letting properties for rent; travel; flying 'simple' aircraft; spending more time with family members (perhaps in recognition that busy rosters and intercontinental travel had served to limit contact with sons and daughters when they were growing up); leading a more ordinary life (as one 60 year old respondent explained: '[I intend to] sleep at night [and] socialise at weekends [and] plan travel beyond this month's roster').

Ultimately, pilots' ability to enjoy their retirement is linked to their own retirement provision, the impact of prevailing economic conditions on their investments and their physical health. For a pilot who leaves the industry as a Captain and who has made adequate provision, retirement should hold no fears. *The Guardian* (2005b) notes of those who plan for retirement: 'Some 80% of the [British] nation's wealth is held by the over-50s .... The top 20% of pensioner couples are prosperous, with an

---

2 At least one respondent knew of someone who had died very soon after retiring (it is possible that the enforced idleness of retirement had been a factor in his sudden demise).

income of £45,000 or more a year'. As for the rest, however: '[B]elow that top band, the other four bands all fall below average income. And at the bottom, there are two million pensioners living below the poverty line ...'.

Investment income is linked to the state of the financial markets. The financial markets are influenced by the state of the world economy. In the 2005 crystal ball publication *The World In 2006*, the staff of *The Economist* attempted to predict the near-term health of the global economy. The picture was far from clear, they said. While they foresaw the possibility of growth in the near-term, things could change very rapidly: 'The exceptional nature of the world economy over the past two years is startling. Global GDP grew at an average rate of 4.7%, its fastest pace in any two consecutive years since the 1970s .... The Cassandras who have long predicted a slump in American consumer spending – one of the main driving forces of global growth – have so far been proved wrong. Yet there are good reasons to believe that 2006 will be the year when consumers in America and several other economies lose their nerve' (Woodall 2005, p. 15).

Economic downturns are bad for both the employed and retirees. The employed can be made redundant. Those who eschew migrant working may join the ranks of the long-term unemployed. Retirees may see their assets devalued and investment income reduced. This is cold comfort for retired pilots. Having survived the vagaries of the aviation industry and possibility of losing their licence through ill health, they may find themselves facing retirement on a much-reduced income. If you add to this the fact that they no longer occupy a position of authority, retired pilots may find it difficult to adjust to their new circumstances. Given that people are living longer, retirees may have to suffer this for a considerable length of time. As *The Guardian* (2005b) has noted: 'Life expectancy has been lengthening for 200 years – by roughly two years a decade'. As to what retirees with money actually do when they are retired, *The Guardian* (2005b) observes: 'We are not just living longer, but enjoying healthier and more active lives too. Ask the travel industry. Older people are signing up in their thousands for activity holidays. Take trekking. Typically 50% of participants are over 50 and 25% over 60 – even for Himalayan treks. It is all part of the new kind of "skiing" (spending the kids' inheritance). Some bank surveys have found six out of 10 older people intending to use some of their capital to support a new lifestyle'. Several respondents planned active retirements. Typical responses were: 'Part-time instruction, boat and aircraft-building. Also sailing and flying my homebuilt ... '; 'Buy a sail boat and sail around the world for a year'; 'We have a smallholding which I look forward to working on'; 'Retire to farming in Scotland with the financial security of a pension'.

## Summary

Most respondents had made plans for either their career or retirement. For those still working some wanted to fly short-haul, others long-haul. A number wanted to become Training Captains. Typical statements were: 'I fly Airbus A321s short-haul.

I want to fly Airbus A330/340s long-haul' and 'I would like to get involved with training to develop my skills further, put something back in, help people along ... enable more varied use of my brain than just "button-pushing" an Airbus between LHR and MAN four times a day, twenty times a week, etc.' Pilots' desire to enhance their skill-set and 'give something back' says much about their sense of responsibility and professional culture. Those facing retirement had made every type of plan, from maintaining some kind of link with aviation (by flying light aircraft, for example) to getting as far away from the activity as possible (by, for example, running a smallholding or sailing the world).

# Chapter 8

# Conclusions

The sociology of commercial flight crew is complex. A significant minority of respondents came from relatively humble beginnings and/or struggled to find a career that interested them. More than fifty percent of respondents had not attended university. Not all respondents had well-paid, secure jobs and glamorous pool-side lifestyles. As the industry adjusts to internal pressures (like more competition from the low-cost sector) and responds to external threats (like new plague-type diseases, natural disasters, terrorism and escalating fuel costs) it is possible that pay and conditions will be squeezed further. It is clear therefore that there is no simple, convenient pilot-stereotype (journalists please take note).

Those respondents and interviewees lucky enough to have senior positions with conscientious employers and to have been in post for a reasonable length of time were – in relation to most of the rest of their country's labour force – reasonably well rewarded. Benefiting from good salaries, they were able to provide for their family's current needs and invest in real-estate with a view to accumulating capital for their family's future needs. They were able to run one or more new cars and take good holidays. They were able to put their children through university and indulge their own interests (by running a boat or classic car or joining a golf club, for example).

Interestingly, not all those pilot-respondents with what most would consider comfortable lifestyles saw themselves as being especially fortunate. Pilots' self-perceptions could surprise, as with this comment from a woman pilot with a five-bedroomed, £600,000[1] main home and two other dwellings (total value £300,000): 'Less young people wanting to be pilots as they see the work getting harder and very tightly regulated ... whilst salaries and packages seem to be getting cut and streamlined, e.g. final pension schemes disappearing, etc'.[2] This respondent's husband was also a pilot. They had no children (and therefore no school or university fees). Another pilot remarked: 'The pressures are greater now. Limits – regarding

1 A 2004 valuation.

2 As mentioned earlier in the book, it is important to distinguish between disposable income (made up of salaries, interest on investments, rental income, etc) and the value of capital assets like houses, flats, cars, boats, etc. It *is* possible to be simultaneously capital-rich and cash-poor. First, because capacious houses with large plots, flats in city centres, boats and prestige or classic cars cost a lot to maintain, insure and run, and secondly because even the most impecunious First Officer can inherit an unfeasibly expensive property or classic car or other item from a deceased relative. (The solution to this trap, of course, is for the inheritor to realise the inheritance at the earliest opportunity, rendering her/himself solvent (for a short while at least).)

flight-time limitations – have become the normal working schedule. The pay is now not so good, especially considering the hours worked. My working hours, when considering the time available to me at home, equated to a sixty-hour week'. This pilot had three children, all of whom were working. The pilot and his working wife lived in a five-bedroomed house valued in 2004 at £700,000.[3] They had a four-bedroomed second or holiday home in the Turks and Caicos Islands valued in 2004 at $395,000. They owned three cars, an Audi A4 Avant, an Audi 80 and a Peugeot 306. A third respondent who claimed that: 'Career rewards seem to be diminishing in both lifestyle and financial reward' lived in a house with a 2004 market value of £700,000 and owned a Jeep Grand Cherokee and Golf VR6. The respondent who stated that: 'Rewards do not justify the financial input required' owned a six-bedroomed house with a market value of £1,000,000.

However pilots see or 'construct' themselves, the fact remains that senior pilots working in advanced First World countries for prestigious carriers can expect to be *relatively* well paid. Regarding UK salary levels in 2004, Summerfield and Gill (2005, p. 70) observe: 'In April 2004, directors and chief executives of major organisations, with median earnings of £1,791 [gross] per week, topped the earnings league for full-time employees .... Senior officials in national government, medical practitioners and aircraft pilots and flight engineers also had median weekly earnings exceeding £1,000 [actually £1,094]. The lowest paid of all adult full-time employees were leisure and theme park attendants and floral arrangers and florists, with median earnings below £200 per week, only just over a tenth of the earnings of the highest paid occupational group'.

Pilots' undoubted enthusiasm for their craft should not be taken to mean that they are completely sanguine about their industry. Far from it. Many pilots expressed concern about, amongst other things, terrorism, flight-time limitations, the impact of the generally successful low-cost model on venerable full-service carriers, industry rationalisation, the ever-more-radical technologisation of the flight deck, the quality of airline management, the effectiveness of intra-company communication and feedback loops, the effectiveness of internal audits and of national and international regulation, pilot deskilling, flight deck ergonomics, and the impact of flying's forced sedentary lifestyle on short and long-term health.

Perhaps the most startling finding, and one that must be of concern, was the fact that a majority of respondents said they would not encourage their offspring to take up a career in commercial aviation. (Thirty-five percent of respondents said they *would* encourage their offspring to become a commercial pilot. Sixty-five percent said they would not.) We need to understand why those who know most about commercial aviation's 'coalface' (line flying) and who are committed to their profession are so reluctant to expose their offspring to it. Could it be that the industry has changed to such a degree that it is no longer considered a worthwhile pursuit? This comment from an ex-RAF Flight Lieutenant working for a far-Eastern airline reflects the concerns of numerous respondents: '[I am concerned about] commercial

3 In December 2004 the average price of a house in the UK was £152,623.

pressure; a management attitude that finance is everything; performance-related pay awards when a human resources desk-jockey tells me how well I am flying my aeroplane, and consequently what my pay-rise will (or will not!) be. The industry has gone crazy'. Another respondent, who remarked '[there is] no respect from upper management with regards quality of life', continued:

> We need better quality engineering. Do not contract-out engineering. Train and use your own engineers; FTLs for long-haul in particular are not realistic; we need more time off between flights and management need to be made aware of the lifestyle of pilots and their families; management need to participate in CRM programmes with pilots, regularly.

While it is true that senior pilots with good jobs at prestigious airlines enjoy high salaries and comfortable lifestyles,[4] it is also true that flying is not the most secure of careers. A failed medical, industry slump or media slur can terminate even the most glittering career. All pilots live with the knowledge that everything they have worked for can be taken away almost in an instant. They also know that their skills are not easily transferable to other occupations. Pilots, therefore, are *vulnerable*. As one 76 year old respondent put it: 'Piloting is a very insecure career as a fail by the operator or of the medical can have drastic results for the individual pilot who normally has no other skill or qualification'.

Pilots' vulnerability is increased by the interactive complexity and close coupling[5] of the industry's constituent parts, and by the interactive complexity and close coupling of the industry with its social, economic and political milieu. Examples of close coupling and interactive complexity abound in commercial aviation. For example:

*Case I*

In August 2005, a catering company based at Heathrow Airport sacked several hundred of its employees. Financially-troubled Gate Gourmet claimed that its workforce was inflexible. Denied its in-flight meals and hit by a sympathy strike by baggage handlers, British Airways cancelled many of its services. Some 70,000 travellers were affected. Passengers interviewed live on television voiced their frustration with the airline. (Because it resonated with the media's news agenda – specifically its appetite for bad news – the dispute received significant coverage.) It was estimated that the cancellations cost British Airways £20–40 million (Baker 2005, p.21) at a

4 As explained in Chapter 5: Pilots' Commercial Flying Career and Perception of the Industry, while some pilots are well rewarded, others – especially those starting out – are not.

5 In a closely coupled system a perturbation in one element impacts proximate elements in a more-or-less direct manner (the constituent elements of a closely coupled system exhibit a high degree of connectivity or mutuality). Systems in which events have difficult-to-predict, multiple outcomes are described as interactively complex – see Perrow's (1984) book *Normal Accidents* for a full definition of the terms.

time when margins were being squeezed by vigorous competition (from LCCs like Ryanair, easyJet and Flybe) and high oil prices.[6] The following statement by two British Airports Authority (BAA) managers alludes to the interactive complexity and close coupling of the aviation industry: 'On the morning of 11 August 2005, simmering industrial tension at ... Gate Gourmet boiled over into a full-scale walkout .... Within a matter of hours, this ... isolated event had triggered a wildcat strike at Heathrow's largest airline operator, British Airways (BA) [BA is Heathrow's largest customer, accounting for over 40% of the airport's traffic], disrupting the travel plans of more than 100,000 people .... *Heathrow is a large operation working within a small ... area with lots of interdependent things happening in a tightly coordinated manner. This means that when something goes wrong, the potential for an incident to escalate quickly ... is very real*' (my emphasis) (Harkin and Hunt 2005, pp. 11–12).

### *Case II*

Just prior to the events at Heathrow, a fledgling Eire-registered low-cost carrier, EUjet, went into administration. The airline's collapse forced the closure of its operating base, Kent International (a former RAF base that had been redeveloped). The airline had commenced flying from Kent International in September 2004. It was hoped that both the airline and airport could be re-launched.

### *Case III*

In the aftermath of the September 2001 terrorist attacks, Switzerland's flag carrier, Swissair, collapsed. Zurich Airport, one of Swissair's main hubs, feared its viability would be affected. Large international carriers like Swissair deliver passengers to hubs where some transfer to LCCs, regional carriers or air taxi companies to be flown to tertiary airports. A single collapse may impact numerous carriers and airports (and local, regional and national economies). Regarding Swissair's collapse, Fleck *et al.* (2001, pp. 6–7) opined: 'Switzerland is not a country that is given to panic. But the collapse of its national airline ... shook the nation to its core and plunged its business community into crisis .... In less than a week the collapse of Swissair inflicted more damage to the reputation of Swiss business than it has seen since the end of the second world war ... '.

---

6 In the Summer of 2005 the Secretary General of the Association of European Airlines (AEA) stated: 'The bottom line has become so sensitive to swings in fuel price that an extra couple of dollars a barrel shows up immediately in the airline's share price. Within this scenario of heightened instability, developing the business in an orderly way becomes more and more difficult' (Schulte-Strathaus cited in Baker and Field 2005, p. 11).

*Case IV*

In June 2004 US regional carrier Independence Air cut itself loose from former partner and major carrier United Airlines, unleashing a chain reaction that eventually forced the 'escapee' into Chapter 11 bankruptcy protection. Rivals reacted to Independence Air's bid for freedom by, amongst other things, matching the airline's fares, reducing ticket restrictions and boosting their advertising budgets. Independence Air buckled under the pressure. Following the airline's demise, low-cost carrier Air Tran said it would fill the void left by the airline's demise by offering more services (Field 2005c, p. 17). This episode demonstrates both the introverted, claustrophobic nature of the industry and its capacity for vigorous (if not vicious?) competition.

*Case V*

Close coupling and interactive complexity can produce surprises. There is, for example, the paradox of the demise of US majors United Airlines, US Airways, Delta Airlines and Northwest in the midst of a consumer spending boom. The link between US consumer spending and the fortunes of its legacy carriers is subtle, but no less deadly for that. During the early years of the New Millennium, US consumer spending was one of the main drivers of global economic growth. As countries like India and China tried to meet rising demand for finished goods, their appetite for raw materials grew. Oil prices rose, exposing the financial and structural weakness of North America's legacy carriers (and also of European legacy carriers like Alitalia). Political instability in the Middle East reinforced the upward pressure on oil prices (Elliott 2004). Northwest's CEO commented: 'We had developed a plan to restructure Northwest outside of Chapter 11 and have been implementing that plan. Unfortunately, in addition to an uncompetitive cost structure, our efforts have been overtaken by sky-rocketing fuel costs' (Steenland cited in *Airliner World* 2005a, p. 16).

The outlook for aviation's original prime movers is bleak, especially if they emerge from Chapter 11 bankruptcy protection with more or less the same business model as that which consigned them there in the first place.[7] Woodall (2005, p. 15) observes: 'With oil demand growing strongly and production and refinery capacity unusually tight, prices are likely to stay high in 2006. Oil prices have more than tripled since late 2001, a rise similar in scale to the price jumps of 1973–1974 [and] 1978–1980 ... '. In the third quarter of 2005 US airlines lost $2.6 billion.

Despite pilots' vulnerability there was clear evidence of the durability of pilots' professional ethic. Many respondents said they found flying rewarding. It was, they said, something that gave them a 'sense of achievement'. As one respondent put it: 'I take pride in having a profession to maintain and in providing a service ... '. Another

7 Tarry (2005, p. 72) says that our fixation with the industry's 'serial loss-makers and underperformers' needs to change.

stated: '[I get] [f]inancial reward. Professional pride. A sense of achievement in a job well done. Team work ... '. A few respondents offered a different construction, however. For example: 'Previously I got camaraderie. Now it is just pay!'

Such views were, however, in the minority. There was strong evidence of a distinct 'pilot culture'. This culture was characterised by, amongst other things, a love of flying, a desire to be 'professional' at all times and to deliver the safest and most efficient service with whatever resources were available (if necessary by drawing on the pilot's personal reservoir of *the right stuff*). One interviewee, asked how he coped with fatigue, simply said: 'You put your head back and get on with it ....You do not enjoy it .... You have to work hard to concentrate on what you are doing ... not to make any stupid mistakes'. In their book *Culture at Work in Aviation and Medicine. National, Organisational and Professional Influences*, Helmreich and Merritt (1998, p. 70) claim that flight crew have a '... shared set of professional beliefs and practices'. In their research into pilots' and flight attendants' 'perceptions of the stresses they encounter at work and in their personal lives' Sullivan and Edman (2004) gathered evidence that '... suggest[ed] the existence of a global supraculture among people who fly for a living'. This 'occupation-based supraculture' transcended 'culture of origin or geographical location of employment'. The data presented here supports Helmreich and Merritt's and Sullivan and Edman's suggestion that there exists a culturally transcendent *community of professional flyers*.

Of course the pilot culture revealed by the research had other, less glamorous dimensions, such as the divorce rate, persistence of sexist attitudes amongst a small minority, and percentage who consume alcohol (ninety-four percent). These traits should give cause for concern. Divorce is an unnerving and (in the UK) financially testing experience that can impact self-esteem and motivation. Low morale is not conducive to good performance at work. People generally have to feel good about themselves to give of their best.

Sexist attitudes harm the industry because they reduce the supply of pilots. Being irrational and morally indefensible they are also a sleight on those who propagate them, and on colleagues who tolerate those who propagate them. Of course aviation is far from a special case. Sexism and discrimination remain a prominent feature of social relations generally. They are manifest in employees' attitudes, recruitment patterns and wages and salaries. Osborne (2005b) notes: 'The [UK] government has been urged to take action after an official report revealed that although the gender pay gap has narrowed, it remains in double figures'. Responding to the report's findings, a spokesperson for the Equal Opportunities Commission stated: 'It is time for government to ask employers to take more proactive steps to address this persistent problem, through the introduction of a requirement on the private sector to promote sex equality and eliminate sex discrimination' (Watson cited in Osborne 2005b).

Given that the culture or 'tone' of an organisation is in part a reflection of the opinions and actions of those at the top, it follows that appointing more women to positions of authority may have the effect of improving women's status. In the UK the number of women in senior positions is still quite small. As *The Guardian*

(2005a) has noted: 'Countless reviews, government initiatives and reports have failed to make any real impact on boardrooms stuffed with what Derek Higgs described as "white males nearing retirement" and what BP's Lord Browne once referred to as the "golf club culture" of UK boardrooms. The average age of a male non-executive ... is still 60'. The industry publication *Airline Business* often carries a picture of an airline CEO on its front cover. Almost all of the featured CEOs are men. While there is no suggestion that any of these CEOs hold sexist attitudes or practice discrimination, men's dominance of commercial aviation management creates an image that may deter some women and feed prejudices in society generally.

The issue of alcohol consumption is less straightforward. Taken in moderation alcohol can have a beneficial impact on health. If taken to excess, however, alcohol gives serious cause for concern (Smither 1988, p. 392). In its 2004 action plan *Alcohol Harm Reduction Strategy for England*, the British government estimated that some 17 million working days were lost due to excessive alcohol consumption and that alcohol misuse cost the economy around £20 billion per annum (Cabinet Office 2004, p. 7). It is interesting to note that with regard to alcohol consumption, pilots are probably no different to any other professional group. As Summerfield and Gill (2005, p. 101) explain: 'Alcohol consumption is higher in managerial and professional households than in routine and manual ones'.

Kohn (2005, p. 27) states: 'Pilots ... need an excellent level of health and fitness .... In consequence there is a need for a commensurate lifestyle to keep fit for the regular yearly health and fitness checks throughout a pilot's career, which become half-yearly after age 40'. The research revealed some positive indicators with regard to pilots' health. Few smoked and most took some form of physical exercise (although those who were in a position to cycle to work chose instead to drive, or take public transport). Given pilots' comments about work-induced fatigue and stress and the fact that piloting is, for the most part, a sedentary activity, pilots' determination to stay fit (by taking exercise and not smoking) indicates a mature attitude. If the latest (2005) research proves correct, those who exercise can look forward to a reasonably active retirement: '[R]ecent research from the US national centre for chronic disease prevention showed lifestyle is more influential than genes in avoiding age-related health problems. We all have a direct influence on how long we live' (*The Guardian* 2005b). Unfortunately, pilots' comments about the availability and quality of food indicates a lack of understanding on the part of some employers about the importance of a balanced, healthy and energy-sustaining diet.[8]

8 Commercial aviation is not the only industry to be slow to understand the importance of healthy eating. In 2004/2005 a terrestrial channel in the UK (ITV Channel Four) engaged a celebrity chef (Jamie Oliver) to investigate the quality of the British school meal. Oliver found that many children were fed a diet of processed, low-quality foodstuffs, biscuits, crisps, chocolate and sugary fizzy drinks. Oliver confronted the British Prime Minister at 10 Downing Street with his findings and a sizeable petition. The Prime Minister promised to act, eventually giving UK schools an extra £235 million. The documentary series received such acclaim that it was repeated in December, 2005. Healthy eating can prove costly, however. Stallwood (2006) notes: 'The high-fat, high-salt food could return – to the delight of schoolchildren

Staying fit and healthy is not only a matter of eating balanced meals and taking exercise, however. It also requires the mitigation of environmental hazards, like polluted air and cosmic ionising radiation. This is not easy. For pilots and cabin crew (and passengers) exposure to such hazards is unavoidable. Hazards like radiation may have serious long-term health impacts on flight and cabin crew. Whelan (2003, p. 805) notes: '[I]n the past 2–3 years, 10 new studies of pilots or cabin crew have been published .... Overall, the reports suggest that pilots are at increased risk of malignant melanoma, non-melanoma skin cancer, and possibly acute myeloid leukaemia, and that cabin crew are at increased risk for breast cancer and malignant melanoma'. Sounding a note of caution, Whelan also says: '[T]he question remains as to whether the observed excesses are due to occupational exposures or non-occupational factors, such as reproductive history and lifestyle'. Clearly, the contribution of such 'confounding factors' as reproductive history to cancer rates amongst flight and cabin crew requires detailed assessment. Regarding the incidence of breast cancer amongst female cabin crew, for example: '[I]nvestigators estimated that the reproductive history pattern in female cabin crew (that is, fewer pregnancies later in life) could yield about a 10% increase in breast cancer incidence compared to the general population ...' (Whelan 2003, pp. 805–806).

Pilots' comments about some employers' interpretation of flight time limitations (FTLs) give cause for concern. This comment was typical: 'Airlines must use FTLs as a guide – not a target'. FTLs were introduced to specify an occasional maximum to be attained only in exceptional circumstances. Flight time maximums were never intended as norms. Regarding the accusation that airlines increasingly frame or construct FTL maximums as targets, the airlines' defence – that they are acting within the law – is unchallengeable. While airlines may not be using FTLs in the right spirit, they are certainly not breaking the law.

While these airlines have no legal case to answer, it *could* be argud that they do have a moral case to answer. Although the morality of using FTLs as rostering targets cannot be debated in a court of law (not in the UK, at least), it could be discussed in the court of public opinion. The media could help in this regard by foregrounding the issue of acute and chronic pilot fatigue (arguably a more constructive use of their time and collective intellect than writing about celebrities, cars, cooking or holiday homes). Airlines that construct FTLs as targets need to be aware that they are out of step with current thinking on working hours and the work–life balance generally. Ironically, while more UK pilots are being required to work to the FTL limit, junior doctors' working hours are (at last) being moderated: 'By 2009, the working hours of [UK] junior doctors will be cut back to 48 hours a week .... There will be an interim restriction of 56 hours in 2007 .... In the past, wards often depended on junior doctors working long hours, sometimes more than 100 hours a week ... (Callaghan 2005)'. According to Bunting (2004) long hours are a source of gender discrimination: '[The

– after the average cost of a school meal rose to £1.85 in one London Borough'. Previously children had been fed for much less than a pound each. Airlines would be aware of the cost differentials.

long hours] work culture is effectively a form of discrimination against women because it becomes impossible to combine a job with long hours, with the caring responsibilities which fall, more often than not, to women. The long hours culture impacts on the labour market at both ends – the poorest paid and some of the best paid – and in both cases, most women get squeezed out'. According to the Trades Union Congress report *Challenging Times*, some 2.3 million UK workers have expressed an interest in working fewer hours (for less pay, of course) (Ward 2005).

In 2004, pilots opposed to the proposed pan-European harmonisation of FTLs won a small victory when: 'Transport ministers from the European Union's 25 member states unexpectedly rejected plans to raise the maximum length of a shift in the cockpit from 12 to 14 hours and to cut rest times between flights .... Airlines were strongly in favour of the proposals .... Balpa cited a study of 55 fatal accidents by the US federal aviation authority [sic] which concluded that crews were six times more likely to crash after 13 hours on the flightdeck' (Clark 2004c). BALPA argued that the scientific case for more liberal FTLs had not been made. '[O]n the 11th June [2004] ... the Council of Ministers are being asked to set new European-wide rules that have less to do with science and more to do with political compromise' wrote BALPA's General Secretary (McAuslan cited in BALPA 2004). Following rejection of the plan, *Air Line Pilot* noted: 'European policymakers have sought to standardise flight time limitations across the European Union for more than a decade now .... The intended law ... would have had a detrimental effect on pilots with more stringent flight time rules' (Francis 2004, p. 34). A key issue, therefore, is whether FTLs are harmonised to their most, or least, liberal manifestation. For the sake of fair competition it is clear that *all* regions should subscribe to the *same* FTL. The question is, on what basis should FTLs be calculated (scientific or political criteria), and by whom?

As in any industry, aviation has its good employers and its mediocre employers. As mentioned above, some managements construct FTLs as targets. Others are more conscientious. Several respondents commented on the manner in which airlines were run. A common theme was the distance between an airline's directors and managers and its pilots. As one respondent put it: 'It is important that company directors have knowledge of the work carried out by their staff – preferably, but not necessarily, they should have piloting experience'. Another commented: 'Companies need to value and listen to their employees'. A third stated: 'Install a management that appreciates that pilots are not idiots working on a production line or labouring on a building site. Pilots like to be informed and included in the management loop'.

Pilots' aspirations for a more knowledgeable, considerate and participative style of management suggest a need for greater workplace democracy. Exactly what form this should take is open to debate. According to Theobald (1994, pp. 120–124) there are two basic forms of participation. The most radical is power-centred participation. The least radical (and the form most often adopted by management) is task-centred participation. Theobald comments: 'Participation can ... run from employees being occasionally consulted about management decisions to the kind of highly institutionalised participation in executive decision-making that exists in some European countries .... Codetermination was adopted in the then Republic of West Germany as part of the restructuring and de-Nazification of German labour relations after the Second World War. The basic principle underlying it is that workers have a right to participate in the running of the firms they work for'.[9] Task-centred participation, unlike codetermination, leaves extant corporate power structures in place. Task-centred participation's most common manifestation is the Quality Circle: '[Originating] in the United States, Quality Circles were exported to Japan and from the early sixties were widely adopted in manufacturing and in some areas of white-collar employment. A Quality Circle comprises a group of people within an organisation who meet together on a regular basis to identify, analyse and solve problems concerning quality, productivity or any other aspects of the day-to-day arrangements of working life'.

Likert's work is also relevant to this debate. According to Smither (1988, pp. 218–219) Likert identified four management styles: *exploitative authoritative* '... in which all decisions are made at the top'; *benevolent authoritative* which '... allows for some upward communication'; *consultative* where '... communication upward is limited, but employees do influence decision-making'; and *participative* where '... communication at all levels is open and honest, and workers have direct influence in making decisions'.

While pilots enjoy a great deal of freedom on the flight deck (where, during the course of a day's flying, they are required to make decisions pertaining to safety and economic efficiency) the *modus operandi* of the airline for which they work would appear to be largely outside their control. The fact that pilots are simultaneously empowered (as professionals) and disempowered (as employees) creates frustration. This frustration is evidenced in pilots' statements. Clearly the industry should consider whether it is making the best use of one of its key (and most expensive) resources – flight crew. Research by Irwin (1995) and Bennett and Shaw (2003) suggests there is much to be gained from exploiting workers' experiential knowledge.

As enterprises become more complex and pressured, management's knowledge of the 'lived experience' of its workforce diminishes. Feedback mechanisms can help restore management's knowledge base. Restoration of the knowledge base ensures that rules and regulations are realistic and practical (i.e. implementable), and that safety margins are maintained. Without consultation, management becomes disconnected (detached). The models that inform management decisions become unreal. Decisions made by management with reference to such models may be

9 The British Army was partly responsible for the codetermination initiative. At the end of the Second World War much of German industry lay in ruins. German society was severely ruptured – a potential breeding ground for far-left or far-right radicalism. Aware of the danger, the British Army set about resurrecting Germany's automotive industry. British Army officers encouraged Volkswagen workers to get involved in the running of the company. Volkswagen thrived. The codetermination movement was launched. German society began to coalesce and stabilise.

ineffective or dangerous. Such decisions may frustrate or alienate workers. They may also alarm regulators and investors, and attract the attention of the media.

Airline managements should seek to avoid alienating their pilot groups. In return Captains and First Officers should seek to make a positive contribution to the company *as an enterprise* – a contribution that balances professional satisfaction and personal reward with the need to keep the airline profitable in the face of increasingly aggressive competition. As one British Prime Minister has remarked, citizens have 'rights *and* responsibilities'.

In recent years the European Union has given workers consultation rights with employers. The UK government has also tried to encourage closer working between employees and managers through initiatives like the multi-million pound Partnership Fund. The TUC's General Secretary welcomed the EU initiative, saying: 'For the first time employees in all Europe's medium and large companies will have the right to find out what their firms plan, and have their views listened to' (Monks cited in Black and Wintour 2001). Results in the UK have been mixed, however. As Sheffield University's Professor Stephen Wood (cited in Ball 2005) explains: '[A]s yet the reforms have not contributed greatly to the competitiveness and innovation agenda'. There have been blockages on both sides of the industrial divide. As Ball (2005) explains:

> Another aim [of the 2001 EU legislation] has been to encourage partnership between employers and unions. The theory is that this offers employees a greater say than old-style collective bargaining; organisations are said to get better performance. Incentives included [the] Partnership Fund, which, until 2003, supported projects to foster closer working. But [Professor] Wood says the jury is still out, with research needed to measure precisely what the glossy brochures have produced in practice. Responses to new employee rights to consultation ... have been decidedly muted. These regulations allow employees to force the establishment of works council-type consultation. One survey suggests firms have been trying to comply, but with more form than substance so far. 'Employers see staff apathy as a big issue', says Mark Crail, editor of IRS Employment Review. 'Many of these bodies seem fixated on dress codes and toilet paper. It is a long way from the strategic engagement expected'. Workers have not been enforcing their rights through legal channels either. Just three applications have been made since April for intervention by the Central Arbitration Committee, which is charged with putting workplace arrangements into effect. The 2004 Workplace Employee Relations survey charts a decline in employee representation, with joint consultative committees in just 14% of workplaces, down from 20% in 1998. All round, Labour's period in office could be seen as a setback for employee involvement and collectivism, despite talk of partnership and engagement.

Crail's point about worker apathy is an interesting one. It is one thing to call for closer working relations with management (as many pilot-respondents did) but quite another to find the time from busy schedules to make it work. Collectivism requires active participation in discussion forums. If participation in local, regional and national elections is any indication of the public's commitment to participative democracy and collective decision-making, the partnership movement has a long way to go – and not only in the UK. As Whiteley (2002) explains:

> There has been a significant drop in the proportion of citizens willing to vote in elections for the European parliament, and in general and local government elections. Turnout has declined significantly in all of these during the past decade .... [Indeed] turnouts in national elections in a wide variety of advanced industrial democracies have declined over the past 40 years .... So the problem is almost universal.

It's not all bad news, however. Writing in 2001, the UK Labour Government's Trade and Industry Secretary noted the success of Citizens' Panels[10] in settling local disputes. 'We should make more use of such forms of "deliberative democracy"' she said (Hewitt 2001). The Secretary of State noted how a Citizen Panel had helped resolve a bitterly disputed rationalisation of health service provision in her constituency city of Leicester. According to Whiteley (2002) '... where individuals are linked together in networks of civic engagement' participation in formal votes (like referenda and national elections) is relatively high.

It was evident from the responses that pilots care deeply about their profession. Responses indicated that pilots take pride in their craft and in standards of flying. Many said they were seeking to give something back to the profession by, for example, getting involved in training. The following responses to the question 'In terms of your flying career, how would you like to see it develop over the next five years?' were typical: 'Get a command. Get involved in training'; 'Long-haul skipper, then training'; 'I want to improve my skills as a trainer while helping others'; 'I want to spend two years in a First World airline, then instruction (three-quarters time) to be with young, bright people and pass something on'. Such responses evidence a sincere concern about professional standards, despite the industry's inherent volatility and pilots' chronic vulnerability.

It is not possible to understand pilots' outlook and behaviour without appreciating their vulnerability. The airline business exists in a state of permanent revolution. The industry's *only* constant is change. Airlines that appear indestructible, like Braniff, Pan Am, Trans World Airlines, Sabena and Swissair, simply disappear. One of the reasons for this volatility is that airlines find it difficult to make money. Petzinger (1995, p. xix) talks about 'The marginal economics of the industry' and 'the proximity of success and failure to every decision'. The airline business is a *knife-edge* business underwritten by Darwin's survival-of-the-fittest theory of evolution. It is a business red in tooth and claw. As Robert Crandall (cited in Petzinger 1995, p. xix) of American Airlines once remarked, commercial aviation is '... intensely, vigorously, bitterly, savagely competitive'. Even Michael O'Leary (cited in Ashcroft 2004), by common agreement a successful CEO, has commented: 'Airlines are very hard to run ... '. But why? Do they not use the very latest technology, pull the slickest marketing tricks, project the most glamorous of images and employ the brightest people?

Airlines find it difficult to make money because they are, in essence, elements of a 'smokestack' industry – an industry that is heavily labour and capital-intensive.

---

10 Citizens' Panels are informal 'juries' convened to debate local policy questions, like health service rationalisation. They are open to members of the public.

Donald Burr, founder of doomed low-cost carrier People Express, understood this fact only too well. He called airlines 'a low-end commodity ... capital-intensive, labour-intensive, fuel-intensive' (Burr cited in Sampson 1984, p. 136).[11] Because of the *modus operandi* of its constituent companies, the airline industry is as vulnerable as any of the old smokestack industries, like ship-building, steel-making or coal-mining. Unlike software companies, investment firms, research institutes, universities and fashion houses, airlines are not, for the most part, low-overhead, lean-and-mean 'knowledge enterprises'. They are expensive-to-run, temperamental behemoths, vulnerable to social, economic and political perturbations. This is the lived reality of the airline CEO and of those who work for her/him. As one American Airlines executive said of his industry: 'Most executives don't have the stomach for this stuff' (Baker cited in Petzinger 1995, p. xix).

### *A reflexive consideration*

It is hoped that this book has contributed something to the development of a sociology of flight crew. In the author's opinion, the industry knows too little about its key resource. While it may be convenient in management terms for an airline to see its pilot group as just another factor of production (like airframes or Avgas) it is economically dysfunctional. A pilot group that is engaged with the airline as an enterprise is a valuable resource.

As explained in Chapter 1, this study is not flawless. Nevertheless, the results obtained in Chapters 3 through 7 correlate with the observations made in Chapter 2. The congruence between the book's primary and secondary data suggests that the observations made here about pilots are broadly correct. Having said this, there is a great deal of work still to be done. More pilots need to be surveyed and interviewed. Cabin crew should also be studied, because navigating an aircraft safely and efficiently between point A and point B is a team-based activity (as confirmed by the scope of CRM training).[12] Maintenance staff, Dispatchers, ramp workers and air traffic controllers are also deserving of sociologists' attention (funders of research, please take note). Commercial aviation is an impressive industry. But it has yet to realise its full potential as a force for good in the world.[13] Quality research can help it do this.

---

11 The paradox of commercial aviation is that it is simultaneously a smokestack industry – labour-intensive and difficult to manage – and a technology-leader. Petzinger (1995, p. xviii) notes: '[Airlines] exploit technology on a scale exceeded by no industry except perhaps medicine'.

12 It has been suggested that the industry dispense with the term cabin crew and refer to all on board as flight crew.

13 ICAO's President has noted: '[Commercial aviation is] one of the most powerful forces to have shaped the evolution of our modern society .... [It is] a powerful force for the continued well-being of every man, woman and child on our small planet' (Kotaite cited in International Civil Aviation Organisation 2003, pp. 12–13).

In an effort to draw the study to a productive close the author has, on the basis of the data presented here, taken the liberty of drafting an agenda of concerns and ideas for the industry's 'movers and shakers'. Chapter 9: An Agenda for Managers, is intended to provoke new thinking amongst those responsible for the industry's regulation and management. The author hopes the reader will forgive this hubristic excursion. The final chapter has been written with honourable intent.

## Chapter 9

# An Agenda for Managers

### Exploit organisational intelligence

It is axiomatic that knowledge and expertise are to be found in all corners of an organisation. Given this fact, managers who fail to exploit this resource are in dereliction of their duty to the company, its shareholders and employees. *They are failing as managers.*

Feedback loops, project teams, problem-solving groups and works councils are all effective ways of exploiting organisational intelligence (or 'distributed intelligence'). Given the tempo of modern business, it is unsurprising that managers and supervisors become isolated. Unable to make the time to 'walk the job' or maintain links with coal-face workers, managers' and supervisors' understanding of issues and opportunities degrades.[1]

Consultation can help restore managers' understanding of the production cycle *as experienced by front-line staff like pilots, cabin crew and check-in staff.* Managers and supervisors who have only a limited understanding of workers' 'lived experience' of the operation *will* mis-manage. Productivity will fall away. More seriously, safety margins may be reduced to the point where latent errors manifest as active errors (see Reason (1995) for an explanation of latent and active errors). Incidents and accidents have direct human costs and can be very bad for business.

There is a *quid pro quo*, of course. Specifically, pilots and other front-line staff must be willing to spare the time to make their contribution to the pot of knowledge. As explained earlier, experiments in industrial democracy have had mixed results (at least in the UK).

### Nurture valuable resources

Donald Burr, founder of People Express, called airlines 'a low-end commodity ... capital-intensive, labour-intensive, fuel-intensive' (Burr cited in Sampson 1984, p. 136). Despite some downward pressure on salaries, pilots remain an expensive factor of production. They are costly to train and employ. If they depart during an industry up-cycle, they can be expensive to replace. Managers may have to offer sizeable

---

1 See, for example, Stanton *et al.*'s (2005, p. 535) case study of the UK railway industry for evidence of 'practical drift' amongst track-workers and signallers.

inducements. It follows that if employee costs are to be kept at a reasonable level, managers should attend to pilots' (reasonable) concerns. Take worries about FTLs, for example. Many respondents and several interviewees believed that FTLs were no longer seen by managers as an occasional, 'last resort' facility. Many believed the upper limits of FTLs were seen as targets or norms. This perception caused great unease. Clearly, if managers are to retain knowledgeable, time-served flight crew (see the point about exploiting organisational/distributed intelligence, above) and keep hiring costs down, they need to listen to pilots' concerns.

Of course, like many issues in aviation, the turnover question is not straightforward. There is evidence that some pilots find the prospect of moving from carrier to carrier (and thereby, perhaps, from continent to continent) attractive. Even the most attentive management would find it difficult to retain 'wanderlust' pilots.

## Exploit talent pools

Despite periodic slumps, the aviation industry continues to expand. Given that the demand for pilots can exceed supply it is vital that the industry taps every potential pool of talent. This means reaching out to groups that are under-represented in the industry. There is evidence to suggest that with more encouragement, better social provision and more flexible working arrangements the industry could attract a greater number of women.

Again, the problem of women's under-representation is not straightforward. While potential female recruits can be deterred by the lack of female role-models and/or persistence of sexist attitudes amongst a small number of pilots, they are also put off by the lack of practical support for working mothers. Only governments can address this latter, crucial issue. The problem of women's under-representation within the pilot-community requires joint action on the part of airlines, national governments, local authorities, educational establishments, trades unions and other public and private agencies. Solving the problem requires 'joined-up' policymaking. Holism is the key.

Efforts to establish systems of 'joined-up government' have focused on the need to improve communication between government departments, agencies, private corporations, special interest groups like trades unions, charities, think-tanks, universities and other centres of power and influence, with a view to implementing a more 'problem-centric' approach to policymaking and budgeting (Bogdanor 2005). Di Maio (2004) explains:

> The term 'joined-up government' was coined several years ago in the United Kingdom ... to describe the integration of services, processes [and] systems ... to achieve a seamless, citizen-centred government.

Clearly the under-representation of women in aviation and consequent waste of talent could be considered one of the many social issues worthy of attention by the kind of problem-solving matrix described by Bogdanor and Di Maio. Addressing the issue of women's under-representation in aviation requires the *effective* co-ordination of stakeholders (such as the airlines, women's groups, trades unions, government departments, educational establishments, local authorities, etc).

## Lobby for an even playing-field

One of the paradoxes of commercial aviation is that while different airlines may use the same equipment (for example, the Boeing 737NG or Airbus A320 family), states' proclivity for safeguarding the national interest (or for being just plain idiosyncratic) means that identical equipment may be used in different ways. Take FTLs, for example. Some states set limits of 900 hours flying time per annum, others 1,000 hours. Airlines registered to low-FTL states are placed at an immediate disadvantage. They may have to employ more pilots (an expensive resource) to deliver the same level of service as competitors registered in states with less restrictive FTLs. Clearly, equity requires that a single (and safe) FTL be applied to all states.

Efforts to harmonise regulations have met with mixed success. While ICAO's SARPS have achieved much, states are still at liberty to go their own way.[2] By striking out on their own, states may be able to gain a significant advantage for their national register airlines. While states' pursuit of the national interest is understandable (at least in the political and economic sense), it encourages airlines registered in states with more restrictive FTLs to treat the flight-time upper limit as an economically necessary target, rather than an 'occasional maximum'. The solution is to set a single, safe and binding FTL for every nation. While this may be considered hopelessly utopian and impractical, it is the only logical answer to current tensions over FTLs. In the matter of FTLs, harmonisation is a precondition of equity.

Harmonisation is very much part of the New Millennium zeitgeist. Despite setbacks (like the 2005 vote by various member states against the EU draft constitution) the underlying trend is for individual states to agglomerate and integrate. Speaking on the future of the European integrative project, Steele (2005) says that the bloc needs '... harmonisation rather than competition. Instead of a race to the bottom, Europe should raise social standards to a common level throughout the union, in maximum hours worked, minimum wages, trade union rights and welfare provision'. Those governments, representative bodies and trades unions who are persuaded of the case for harmonisation should seize the moment and press their case – in as many forums as will give them the time. *Carpe diem.*

2 Everard (2006, p. 17) comments: '[S]upervisory and co-ordinating bodies (national and international) need more muscle in their efforts to oversee flight safety and their approach needs to be more hands-on (including the close monitoring of delegated responsibilities)'.

## Anticipate the media

Since the 1960s, western societies have witnessed the 'decline of deference'. In her book *Fear: A Cultural History*, Bourke (2005, p. 318) describes how the erosion of social deference has impacted the medical profession: 'From the 1960s confidence in [medical professionals] was crumbling fast. In part this was a result of the more diffuse distrust of authority which characterised these years. The environmentalism and anti-militarism of the Vietnam War era spawned doubts about authorities in general .... Scientists and physicians were increasingly reviled in popular culture'. Distrust of the medical profession may have intensified in recent years (witness, for example, the impact on the British medical profession's image of the removal without parental consent of the organs of dead infants).

As for medicine, so too for aviation. Airlines, airport authorities, manufacturers, leasing companies, insurers, regulators, politicians, trades unions and others linked to the aviation industry are considered fair game by the media. Today's 24-hour media industry has a voracious appetite for news – especially bad news, because bad news, like glamour or celebrity, sells newspapers, magazines and air time. Bad news is good news for media companies. Bad news is money in the bank. Bad news is a rising share price and big bonuses for media moguls. In her book *Fear: A Cultural History*, Bourke (2005, p. 321) speaks of the media's 'obsession with death'. She notes:

> With the vast expansion in the mass media during the twentieth century, dangers were publicised to an unprecedented degree, *making them seem almost boundless*. (Bourke 2005, p. 350, my emphasis).

The media is Janus-faced. Some reporting, like that on security lapses at airports, undoubtedly serves the public interest (although only if the story is handled in a sensitive and constructive manner). However, some reporting, like the media's often hyperbolic post-accident speculation, is just plain destructive. Speculation serves no useful social purpose. It confuses, upsets and angers the relatives and friends of those involved in the accident, and frightens the public at large. It harms the airline involved in the accident, and the reputation of the industry as a whole. Everyone is scarred to some degree. Were it not for the relatively inelastic demand for the industry's products, irresponsible reporting could have serious, long-term consequences.

The lesson for those who work in and care about the industry is simple: be vigilant; raise the safety bar wherever you can; perform to the best of your abilities;[3]

---

3 Eyre (1984, p. 174) draws our attention to the fragility of reputation: 'Goodwill and a good reputation are not easily acquired; they take a good deal of time and trouble to build up. Conversely they are very easily and quickly destroyed and once gone take a much longer time to rebuild. Confidence, once lost, is not easily regained. Even relatively minor incidents ... can mar the organisation's public image. The need for special attention to public relations is, therefore, obvious'.

be prepared to meet whatever challenge is thrown up by the media. Crucially, do not get drawn into speculation. Speculation may satisfy the media's requirement for quick and easy answers to complex questions,[4] but it undermines investigators' efforts to deal with the accident in a rational and objective manner ... and discover the truth.

## Publicise and propagandise

If an industry is to prosper it must endeavour to establish and maintain good public relations. For many years the British nuclear and petrochemicals industries took public and media support for granted. Despite heralding *The Chemical Age* and promising 'electricity too cheap to meter', by the 1970s both industries had lost political ground to the noisily vigorous environmental movement. Environmentalists' eye-catching protests and carefully cultivated media contacts had consistently wrong-footed an arrogant industrial establishment. The launch of the BBC drama series *Doomwatch* (which focused on issues of environmental disamenity) confirmed that the environmental movement had all but captured the media agenda. Smarting from largely self-inflicted wounds, the nuclear and petrochemicals industries invested significant sums of money in propaganda and more durable public-relations initiatives, such as visitors centres (the British Nuclear Fuels Limited (BNFL) Sellafield and BP Chemicals Baglan Bay visitors centres proved highly popular).[5] Eyre (1984, p.174) observes:

> Fostering good public relations now plays a very important role both in the private and in the public sectors. It has been described by the Institute of Public Relations as the 'deliberate, planned and sustained effort to establish and maintain mutual understanding between an organisation and its public.' ... [P]ublic relations is charged with more than simply protecting the organisation's reputation. It is also responsible for informing the public about the organisation's actions and policies where they will be likely to affect the physical and social environment. The rise of numerous pressure groups which raise objections to so many plans and projects emphasises the need for careful and expert public relations work to pacify these objectors, or to meet their objections with suitable compromises.

The commercial aviation industry in all its manifestations (scheduled and IT carriers, cargo and crop-dusting operations, corporate outfits, air taxi companies, etc) is *vital* to global economic and cultural development. Aviation drives innovation, supports wealth creation and international trade, facilitates cultural exchange and

---

4 See, for example, Bennett's (2001) review of accidents that had more complex causes than was apparent at first sight.

5 See Bennett (1996) for a review of the BP Baglan Bay public relations drive. BP had been accused of releasing cancer-causing chemicals into the local environment. Residents had cultivated the local and national press, leading to some lurid reporting and putting BP on the back foot.

international understanding, and boosts disaster relief efforts. It creates economic opportunity. It saves lives. It brings a touch of glamour to an otherwise hum-drum existence. It does all this through the efforts of pilots (and numerous other professionals, of course) most of whom never experience the glamorous pool-side lifestyle of long-haul flying. The opening statement of the 2004 drama-documentary *A Wounded Bird*[6] (which told the story of the crash of ASA Flight 529) perfectly captured the essence of today's flight operations:

> Flying isn't always glamorous. Regional airlines are like the buses of the air, trundling back and forth from one small city to another. Crews operate several flights a day, working for up to fifteen hours. They face interminable ground delays, and often surly passengers. If a flight gets cancelled, they don't get paid.

If the industry is to get a fair hearing it must disabuse the general public, their political representatives and the mass media of any false impressions they may have. The industry must also play up its positives, like pilots' enthusiasm and dedication. When ASA Flight 529 came down in a field after suffering a catastrophic engine failure, its pilots were trapped by fire and wreckage. The First Officer, Matt Warmerdam, escaped, but was badly burned. Warmerdam's convalescence was prolonged and painful. Regardless, he was determined to pilot an aircraft again. In *A Wounded Bird* he explained why he was so determined to returned to the line:

> The best thing I ever could have done for myself was that day two years ago when I finished training and took the controls of an ASA plane and flew again. I stubbornly recaptured my dream. And now that I'm doing it again, it has been a joy. It's what I do. It's what I love. It's what I always wanted to do with my life, and I'm doing it again.

Only if the industry broadcasts its positives (like pilots' quiet heroism and the industry's impact on cultural exchange and mutual understanding) can it hope to retain the confidence and co-operation of publics and politicians. Only if it educates the media can it hope for a fair hearing (and even then the media may act on its own skewed agenda). If the industry implements the action points outlined here, *the next one hundred years could be as dazzling as the last.*

---

6 This drama-documentary was one of the programmes in the Aircrash Investigations series produced by Cineflix-Galaxie-NextFilm (a Franco-Canadian consortium). It was broadcast on the UK's Channel Five terrestrial station on 22 December, 2005.

## Appendix 1

# Questionnaire

**Introductory letter**

dd/month/year

Dear Captain/First Officer

**Re: Flight crew research**

Thank you for helping me with my research. I have been commissioned to write a book about pilots and their work. It is part of my project to develop a 'sociology of aviation'.

While there have been numerous psychological studies of pilots and their work, there have been few sociological studies. This book will redress the balance.

All those who reviewed my book proposal (pilots, operations staff, airline managers and regulators) said that, by broadening our understanding of flight crew it would make a valuable contribution to aviation safety.

The book will be written in an open and accessible style. It will be read by pilots, airline managers, industry regulators, students of aviation, journalists and the general public. It will provide for a fuller understanding of pilots and their work and will canvas pilots' opinions on a range of topics. It will give them a voice. It will give them a big audience.

Respondents cannot be identified from the questionnaire. The survey is anonymous. The information will be used to make a positive contribution to the management and safe operation of the world's largest and most important industry. Thank you for your help.

Yours sincerely,

Dr Simon Bennett MRAeS
*University of Leicester, England*

**Questionnaire survey**

Codes [ ][ ][ ][ ]

* **Please write clearly**
* **Please continue on a separate sheet if necessary (citing code(s) (see above) and question number)**
* **Information will be used for academic purposes only. All data will be anonymised**

1 Date of Birth (yy) [ ]

| | Name of village/town/city | Country |
|---|---|---|
| 2 Place of Birth | | |

3 Father's occupation on birth day (if known) [ ]

4 Mother's occupation on birth day (if known) [ ]

5 Did you attend nursery school (pre-school/kindergarten) (y/n)? [ ]

**<u>Education</u> (since age 11)**

6 Please list below your education since age eleven:

| School/College/ University | From (yy) | To (yy) | College/University qualifications |
|---|---|---|---|
| | | | |
| | | | |
| | | | |
| | | | |
| | | | |
| | | | |

If you attended University:

7 Father's occupation on day you commenced? [ ]

8 Mother's occupation on day you commenced? [ ]

9 Did you fly with a University Air Squadron (y/n)? [ ]

**Employment**

10 Employment history <u>prior</u> to becoming a pilot (please note that vacation/part-time jobs should not be listed):

| Employer | Job title | Reason for leaving |
|---|---|---|
| | | |
| | | |
| | | |
| | | |
| | | |
| | | |

Flying career

Military:

11 Have you ever done any military flying (y/n)? [ ]

12 If 'y', for what service(s) (Army, Navy, Air Force)? [ ]

13 What rank did you achieve? [ ]

Commercial:

14 Please list below *all* your commercial flying positions ... from your first flying job to your current flying job:

| Airline | From (yy) | To (yy) | Base(s) | Place of permanent abode (name of village, etc.) | Place of temporary abode (if appropriate) | Reason for leaving |
|---|---|---|---|---|---|---|
| | | | | | | |
| | | | | | | |
| | | | | | | |
| | | | | | | |
| | | | | | | |
| | | | | | | |
| | | | | | | |
| | | | | | | |
| | | | | | | |

## Current family circumstances

15 Parents

| | Alive (y/n)? | Age | Place of permanent abode (name of village, etc.) | Accommodation circumstances (e.g. living with offspring, own home, old peoples' home, hospital, etc.) |
|---|---|---|---|---|
| Father | | | | |
| Mother | | | | |

16 Siblings

| | Alive (y/n)? | Age | Attended Uni (y/n)? | Current job ('H' for homemaker) | Place of permanent abode (name of village, etc.) | Married (y/n)? | Co-habiting (y/n)? |
|---|---|---|---|---|---|---|---|
| Brother #1 | | | | | | | |
| Brother #2 | | | | | | | |
| Brother #3 | | | | | | | |
| Brother #4 | | | | | | | |
| | | | | | | | |
| Sister #1 | | | | | | | |
| Sister #2 | | | | | | | |
| Sister #3 | | | | | | | |
| Sister #4 | | | | | | | |

17 Relationship

Current partner:

| | Age | Attended University (y/n)? | Current job ('H' for homemaker) | Previously married (y/n)? |
|---|---|---|---|---|
| Husband | | | | |
| Wife | | | | |
| Live-in partner | | | | |
| Non-cohabiting boyfriend/ girlfriend | | | | |

If you are divorced, please answer the following three questions:

17.1 Do you believe that rostering patterns played a part in the break-up (y/n)? [ ]

17.2 If 'y', in what way(s) do you think rostering impacted your marriage?

---

17.3 What was your wife's/husband's occupation when your marriage broke-down?
[ ]

18 Offspring

| | Age | Attended Uni (y/n)? | Current job ('H' for homemaker) | Married (y/n)? | Number of children? |
|---|---|---|---|---|---|
| Son #1 | | | | | |
| Son #2 | | | | | |
| Son #3 | | | | | |
| Son #4 | | | | | |
| | | | | | |
| Daughter #1 | | | | | |
| Daughter #2 | | | | | |
| Daughter #3 | | | | | |
| Daughter #4 | | | | | |

**Lifestyle**

Health

19 Do you smoke (y/n)? [ ]

20 If 'y', how many cigarettes per day? [ ]

21 Have you ever voluntarily discussed your cigarette smoking habit with a qualified medical practitioner (y/n)? [ ]

22 What cigarette brand(s) do you smoke?

---

23 Do you consume alcohol (y/n)? [ ]

24 If 'y' have you ever voluntarily discussed your alcohol consumption with a qualified medical practitioner (y/n)? [ ]

25 Do you take exercise (y/n)? [ ]

26 If 'y', what does the exercise consist of and how often do you do it?

---

Residential

27 Please list below your home owning/renting history:

| | Name of Village/ Town/City | Country | House/ flat (H/F)? | Owned/ rented (O/R)? | No. of bedrooms | Approx. value (please state £/$) |
|---|---|---|---|---|---|---|
| Current abode | | | | | | |
| | | | | | | |
| Previous abode | | | | | | |
| | | | | | | |
| Second home (i.e. 'holiday home') | | | | | | |

**Transport/commuting**

28 How many cars do you own? [ ]

29 Please list makes and models:
#1 ..............................................
#2 ..............................................
#3 ..............................................

30 Do you live in temporary accommodation when flying (y/n)? [ ]

31 If 'y', what does this consist of (e.g. bed and breakfast; shared house; rented flat)?

---

32 How would you describe the amenities at your temporary accommodation (please tick appropriate box)?

Basic (utilitarian) [ ]
Comfortable [ ]
Luxurious [ ]

33 How do you normally travel to your base (place of work)?

---

34 How long does this journey normally take (mins.)?

35 If you drive to work, have you ever left your car at work after a day's flying (y/n)? [ ]

36 If 'y', why did you do this and how did you get home?

**Work (please continue on another sheet if you wish (citing code(s) (see boxes at top of questionnaire) and question number))**

37 Apart from flying what other duties do you perform for your airline?

38 What personal rewards do you get from these other duties?

39 In your opinion, are you allowed enough time to perform these other duties to *your* satisfaction?

40 What personal rewards do you get from flying (e.g. financial security; sense of achievement at the end of the flying day; camaraderie/esprit-de-corps, etc.)?

41 What three things worry you most about commercial flying today?

42 What solutions would you recommend for these problems?

43 How do you see commercial aviation developing over the next five years – that is, what do *you* think the major trends will be?

44 In terms of your flying career, how would you like to see it develop over the next five years?

45 What plans, if any, do you have for your retirement?

46 Would you encourage your children to become a commercial pilot (y/n)? [ ]

47 If 'y' or 'n' please give reason(s)

---

48 If 'n', what career(s) would you recommend to them?

#1 ..............................................
#2 ..............................................
#3 ..............................................

49 Why would you recommend these careers over others?

---

50 Please discuss below *any* matter(s) you would like to bring to the attention of the researcher that have not been covered in the preceding questions ...

## Appendix 2

# Interview Questions

Rank and Name [ ]
Interview date [ ]

1 Career choice

1.1 Why did you become a commercial pilot?
1.2 If you had not become a commercial pilot, what career would you have followed?
1.3 Would you describe commercial flying as a job or a vocation (Concise Oxford: 'employment requiring dedication')?

2 Rostering

2.1 Have you ever told Operations that, in your opinion, you are 'too fatigued to fly' (y/n)?
2.2 If 'y', how did Operations react?
2.3 If 'y', can you describe one or more of these episodes?
2.4 Has your partner/wife ever discussed rostering with you (y/n)?
2.5 If 'y', why?
2.6 If 'y', what action did you take?
2.7 Has an adverse rostering experience ever prompted you to consider giving up commercial flying (y/n)?
2.8 Can you describe this experience?
2.9 Have you ever been involved in a pilot fatigue-related incident while flying (y/n)?
2.10 Can you describe this incident?

3 Turnarounds

3.1 LCC pilots
3.1.1 Have you ever worked for a low-cost carrier (y/n)?
3.1.2 If 'y', which ones?
3.1.3 Were turnaround times an issue for flight crew (y/n)?
3.1.4 If 'y', can you describe the issues associated with TT?
3.1.5 How were these issues resolved?

3.2 'Full-service' pilots
3.2.1 When working for 'full service' operators, were TTs ever an issue (y/n)?
3.2.2 If 'y', can you describe the issues associated with TTs?
3.2.3 How were these issues resolved?

4 Rest breaks

4.1 When on duty are you able to take a break during the flying day (y/n)?
4.2 If 'y', what form does the break take?
4.3 Do you feel adequately rested after the break (y/n)?
4.4 (Smokers) Can you smoke during the break (y/n)?

5 Food

5.1 In your current job what access to food do you have during the flying day?
5.2 Do you have any issues with the variety and quality of the food (y/n)?
5.3 What are these issues?
5.4 How would you resolve them?
5.5 When on duty what is the longest you have gone without a meal?

6 CRM

6.1 When and under what circumstances did you first hear the term CRM?
6.2 Did you receive formal training in CRM (y/n)?
6.3 What form did this training take?
6.4 In what way(s) do you think CRM contributes to flight safety?
6.5 What is the worst teamworking experience you have had with a flight-deck colleague?
6.6 Are there any other aviation-related activities that, in your opinion, could benefit from the application of CRM?
6.7 Can you think of any negative aspects to CRM?

7 Ground-handling

7.1 Do you think standards of ground-handling are improving (y/n)?
7.2 What is the worst ground-handling experience you have had?
7.3 How do you think standards of ground-handling could be driven up?

8 Accommodation

8.1 Have you ever lived in temporary accommodation when flying (y/n)?
8.2 What were the pros and cons of living in temporary accommodation near your base?

9 Gender

9.1 Questions to males
9.1.1 Have you ever flown with a female pilot (y/n)?
9.1.2 If 'y', do you think you behaved differently towards your female flight-deck colleague on grounds of gender (y/n)?
9.1.3 If 'y', in what way(s) do you think your behaviour changed?

9.2 Questions to females
9.2.1 Have you ever had a negative reaction from a male flight-deck colleague that you believe was due to gender (y/n)?
9.2.2 If 'y', describe the circumstances and form of this negative reaction.

Appendix 3

# Pilots' Socio-economic Origins and Motivations

## Introduction

These data, derived from the questionnaire, concern respondents' family backgrounds, original social status and career paths to the point at which they became commercial airline pilots.

## Family background

For the purpose of this analysis, class is defined in terms of occupation. Thirteen percent were born into a family whose father had an upper middle-class occupation (for example, a surgeon or company director). Sixty-one percent were born into a family whose father had a middle-class ('white-collar') occupation (for example, a bank clerk). Twenty-six percent were born into a family whose father had a working-class ('blue-collar') occupation (for example, a carpenter, boilermaker or roadsweeper).

Fifty-eight percent were born into a family where the mother had no employment outside the home (that is, she was a 'homemaker'). Of those mothers who were employed outside the home, fifty-two percent had middle-class occupations and forty-eight percent had working-class occupations (for example, factory worker). Forty-two percent were born into a family where both the mother and father had middle-class occupations. Twenty-one percent were born into a family where both the mother and father had working-class occupations. Twenty-three percent were born into a family where the father had a middle-class occupation and the mother a working-class occupation. Five percent only were born into a family where the father had a working-class occupation and the mother a middle-class occupation.

Eleven percent of respondents had no siblings, forty-five percent had one sibling, twenty-six percent had two siblings, seven percent had three siblings, eight percent had four siblings and three percent had five siblings.

Sixteen percent of families where the father had a middle-class occupation were single-child families. Forty-six percent of families where the father had a middle-class occupation had two children. Twenty-two percent of families where the father had a middle-class occupation had three children. Two percent of families where the father had a middle-class occupation had four children.

Seven percent of families where the father had a working-class occupation were single-child families. Forty-one percent of families where the father had a working-class occupation had two children. Thirty percent of families where the father had a working-class occupation had three children. Nineteen percent of families where the father had a working-class occupation had four children.

## Education

Fifty-eight percent of respondents attended nursery school (pre-school or kindergarten). In the case of respondents who came from families whose father had a middle-class occupation sixty-seven percent attended nursery school. In the case of respondents who came from families whose father had a working-class occupation twenty-six percent attended nursery school.

In the case of respondents born in the UK, fifty-three percent attended nursery school. In the case of those not born in the UK sixty-seven percent attended nursery school.

Forty-five percent attended university – although this does not necessarily mean that they obtained a degree. Also, some respondents attended university after qualifying to fly commercial aircraft. Of those resident in the UK several took degrees at the Open University (which offers primarily distance-learning-based foundation and degree courses).

Forty-seven percent of those born in the UK attended university. Forty-two percent of those born outside the UK attended university.

Of those who attended nursery fifty-two percent went on to study at university. Of those who did not attend nursery thirty-six percent went on to study at university.

Of those born into a household where the father had a working-class occupation thirty-three percent went on to attend university. Of those born into a household where the father had a middle-class occupation forty-nine percent went on to attend university. Of those born into a household where the father had an upper-middle-class occupation sixty-two percent went on to attend university (but, as mentioned above, university attendance did not necessarily produce a degree, there being an attrition rate amongst undergraduates).

## Parents' social mobility and longevity

At the date their offspring (i.e. offspring-respondent) attended university, eighty-five percent of fathers inhabited the same occupational class as they had done when their child was born. Five percent of fathers who had a working-class occupation when their child was born had a middle-class occupation when s/he attended university. None of these fathers had moved into an upper-middle-class occupation. Of those with middle-class occupations at the birth of their offspring, none had moved into working-class occupations by the time their son or daughter attended university. Of those fathers whose offspring attended university, sixty-four percent had a middle-class occupation when their offspring was born and when s/he attended university.

At the date their offspring (i.e. offspring-respondent) attended university, eighty-three percent of mothers inhabited the same occupational class as they had done when their child was born. Each of the seventeen percent who inhabited a different occupational class had moved from working-class to middle-class occupations. Of those mothers whose offspring attended university, fifty-six percent had a middle-class occupation when their offspring was born and when s/he attended university.

Of those fathers who had working-class occupations when their offspring-respondent was born, forty-one percent were still alive at the time the questionnaire was completed. Of those fathers who had middle-class occupations when their offspring-respondent was born, sixty percent were still alive at the time the questionnaire was completed. Of those fathers who had upper-middle-class occupations when their offspring-respondent was born, sixty-two percent were still alive at the time the questionnaire was completed.

Of those mothers who had working-class occupations when their offspring-respondent was born, eighty-one percent were still alive at the time the questionnaire was completed. Of those mothers who had middle-class occupations when their offspring-respondent was born, ninety-one percent were still alive at the time the questionnaire was completed.

## Military flying

Twelve percent of respondents had flown with a university air squadron (UAS). Sixty-nine percent of those who had flown with a UAS had gone on to fly for a military service (either Army, Navy or Air Force). Overall, thirty-nine percent of respondents had flown for the military before becoming a commercial pilot. Forty-five percent of respondents born in the UK had flown for the military.[1] Twenty-four percent of respondents not born in the UK had flown for the military. Forty-six percent of those who had flown for the military had also attended university (either before or after their military service). Forty-five percent of those who were born in the UK and who had flown for the military had also attended university (either before or after their military service). Fifty percent of those who were born outside the UK and who had flown for the military had also attended university.

1 One ex-RAF pilot commented that as successive governments reduced the size of the Royal Air Force the percentage of ex-military pilots in the UK commercial fleet would decline. It might be assumed that the same dynamic would obtain in any country with a shrinking air force budget.

## Appendix 4

# Pilots' Lifestyle and Socio-economic Status

### Introduction

These data, derived from the questionnaire, concern respondents' personal lives, their family lives (if any), their wealth, health and travel habits.

### Respondents' relationships and family

Ninety-four percent of respondents said they were in a relationship. Overall forty-two percent of respondents' partners were attending/had attended university. In twenty-one percent of relationships both partners had/were attending university.

Thirty-eight percent of respondents' partners worked as homemakers. Sixty-two percent of respondents' partners had paid employment. Of these, twenty-nine percent worked in the commercial aviation sector. Seventy-one percent did non-aviation related work (teaching, for example). Of those partners who worked in commercial aviation sixty-one percent were cabin attendants, twenty-two percent were pilots, eleven percent were aircraft dispatchers and six percent worked as Inspectors for aviation regulators. Of those partners who worked as pilots, fifty percent had been previously married. Of those partners who worked as cabin attendants, thirty-six percent had been previously married. Of those partners who worked as pilots, seventy-five percent were attending/had attended university. Of those partners who worked as cabin attendants, twenty-seven percent were attending/had attended university. Of those partners who worked as aircraft dispatchers, fifty percent were attending/had attended university.

Fifty-two percent of respondents' partners who did non-aviation work were attending/had attended university. Of those partners who did non-aviation work, eighteen percent had been previously married.

Thirty-four percent of respondents' partners who worked as homemakers were attending/had attended university. Thirty-four percent of those who worked as homemakers had been previously married.

Thirty-five percent of respondents had no offspring. Eleven percent had one child. Forty percent had two children. Twelve percent had three children. Two percent had four children. Respondents' maximum family size was four children.

Thirty-four percent of respondents had children of university age. Seventy-five percent of these respondents had put/were putting at least one child through university. Fifty-two percent of these respondents had put/were putting all their children through university. One respondent had put/was putting all four of his children through university.[1] Analysing child university attendance by place of birth, of those respondents born in the UK with children of university age, eighty-three percent had put/were putting at least one child through university. Of those respondents born outside the UK with children of university age, sixty-two percent had put/were putting at least one child through university.

Of the forty-five percent of respondents who had attended/were attending university, fifteen percent were sending/had sent at least one child to university. Of the fifty-five percent who had not attended university, thirty-five percent were sending/had sent at least one child to university.

The questionnaire offered the chance for respondents to add whatever narration they wished (on the assumption that the questionnaire would not have covered every social and economic dimension of a pilot's life). The following comments were appended in relation to personal circumstances and relationships:

> Once I have driven 1,200 miles a month (minimum), flown 85 hours (average) plus the rest of the duty-time, kept up with all the company paperwork and checks, done two lots of housework and the shopping, put in enough hours of Open University study (to be able to get a real job), kept up to date with Eastenders, oh and slept ..... where on earth do I have time for a relationship?

> Whilst I have worked for only one UK-registered airline and been based at Birmingham throughout, I have worked from most airports in Britain and Ireland, spending protracted periods away from home and accommodated in hotels near the various airports. This is comfortable but a bit soulless!

> I have changed airline jobs in the past in order to ensure a family lifestyle. My Air UK job ended up in permanent night-stops. Within a week of putting in my notice I heard of two more divorces at the airline. A colleague of mine once stated: 'In this occupation you have to keep on changing jobs to ensure a decent lifestyle'.

> Life 'on the road' is tough. Over half a year is spent sleeping in hotel beds. When something goes wrong at home (sick spouse/children, financial worries, etc.) it always seems to happen when you are away. This is a great career, and I enjoy it immensely, but I had no idea about how the lifestyle would affect my personal life. Chasing jobs around the south-west [this respondent flew for SkyWest] I was often in a new city, no friends for good times or bad, and with a schedule that prevented planting roots. You asked about failed marriages. I have never been married but I was engaged once and my career choice definitely played a part in us not getting married. It actually ended up being a good thing. However it does illustrate how difficult it is maintaining a healthy relationship when you

1 In the UK and certain other countries this is a significant financial undertaking given that university fees and living expenses are paid via long-term loans to students. University fees and living expenses used to be covered by local authority grants to students.

are away from home 3–4 nights each week. Your partner must understand not only your job and the nature of it, but also be secure with the relationship as a whole. If I had known from the beginning how difficult the road ahead was going to be it would have given me pause. I probably would have made the same choice, but at least I would have been warned. Working as an airline pilot is, at its best, the most enjoyable, and frankly the 'coolest' job I can imagine. At its worst it is lonely, tiring, stressful and sometimes, when the chips are down and Mother Nature is seemingly angry just at you, frightening.

My departure from Cathay Pacific was prompted by the death of my first wife (from whom I was separated) and need to take care of my children at a critical time in their lives. While at Cathay I had travelled back to Europe for a weekend visit with them every month, which was extremely tiring on top of my regular long-haul flying (of which I had nine years) [this respondent started flying for Cathay out of Hong Kong at the age of forty-two].

Had I been born twenty years later I would likely have had a better education and would probably not have been a pilot [this respondent, born in 1948, went to RAF College Cranwell after gaining two A-levels]. However this would have been a loss for me in career terms. Since I have been an ex-patriate for a large part of my adult life it is difficult to decide whether that or my aviation career has been the main contributor to my divorce (other than my personality) [this respondent had flown for Cathay out of Hong Kong for 14 years].

My wife would like to get pregnant but I am not at home at the right time!

## Respondents' wealth

Respondents were asked to estimate the value of their main property and any holiday home they might own. Thirty-four UK-resident respondents were then selected at random and their worth in terms of the value of their properties calculated. The value of property owned by each respondent averaged out at £372,647 (at 2003/2004 prices). The cheapest property owned was a flat in the Aberdeen area valued at £60,000. The most expensive property was a house in the South of England valued at £850,000. At the end of 2004 the average price of a house in the UK was £152,623.

Forty-four respondents were selected at random and the average number of cars per household calculated. Each household owned 1.85 cars. One respondent owned no car and another five cars.

As mentioned earlier in the book, the value of real estate owned is not the most accurate guide to solvency. It is important to distinguish between disposable income (made up of salaries, interest on investments, rental income, etc) and the value of capital assets like houses, flats, cars, boats, etc. It *is* possible to be simultaneously capital-rich and cash-poor. First, because large houses, flats in city centres, boats and prestige or classic cars cost a lot to maintain, insure and run, and secondly because even the most impecunious First Officer can inherit an unfeasibly expensive property or other item from a deceased relative, making her/him capital rich but without the means to maintain their valuable new asset. (Of course the inheritor can escape from this latter trap by realising the inheritance at the earliest opportunity.)

## Respondents' health

Eight percent of respondents said they smoked. Of these, eighty-eight percent smoked cigarettes and twelve percent cigars. Of those who smoked cigarettes, forty-three percent smoked ten or more per day and fifty-seven percent smoked less than ten per day (in the UK the smallest number of cigarettes that can be purchased is a packet of ten). The most cigarettes smoked by a respondent was twenty per day and the fewest two per day. Sixty-three percent of smokers were in their 40s and 50s. Thirty-seven percent were in their 20s and 30s. Eighty-eight percent of smokers took exercise. Overall ninety percent of respondents took exercise. (Note that respondents interpreted the term 'exercise' in a number of ways. For example, one respondent who ticked the 'yes' box said they went 'Walking, 3–4 times a week, 5–6 miles each session'. Another said they went 'Running, weight-training, climbing ... Usually only 2 or 3 times per week due to jetlag and night-flying. On leave, 5 or 6 days/week'. Another said they went 'Sailboat racing approx. every couple of weeks, walking approx. 1 mile per day'. Another stated they exercised 'Six days a week, either cycling or in the gym. Less regular activities: rock-climbing and skiing [before becoming a pilot this respondent had been an outward-bound instructor]'). Two percent of respondents said they bicycled to work. One pilot who lived in Kegworth and who flew out of Nottingham East Midlands Airport was asked why he did not cycle to work (the village is one and a half miles from the terminal building). He said that in his opinion this was impractical as he would have to wear his uniform and carry a flight bag. His airline provided no changing facilities at Nottingham East Midlands (although there was a gents' toilet on the same floor as his crew room).

Ninety-four percent said they consumed alcohol. Of those who consumed alcohol fourteen percent said they had voluntarily discussed their alcohol consumption with their doctor. (Regarding this finding it should be noted that some doctors may have quizzed respondents on their alcohol consumption thereby drawing them in to an exchange.) One hundred percent of those who smoked also consumed alcohol.

## Travel-to-work

Three percent of respondents did not own a car. All non-car owners travelled to work by bus. All non-car owners lived and worked in the Far East. Of those flight crew who owned cars, thirty-seven percent owned one car, fifty-one percent owned two cars, ten percent owned three cars, one percent owned four cars and one percent five cars.

Eighty-four percent of respondents travelled to work by car, three percent by motorcycle, two percent by bicycle, three percent by train, four percent by bus and four percent by air. The average length of a car journey to work was thirty-six minutes, the average length of a motorcycle journey thirty-two minutes, the average length of a bicycle journey twenty-six minutes, the average length of a train journey forty-four minutes, the average length of a bus journey thirty-six minutes and the average length of an air journey 240 minutes (note that the air journeys were probably made to reach temporary accommodation located close to the pilot's base). Some respondents used more than one mode of transport, as with a pilot based at Heathrow who stated: 'I have a fifty-minute drive to a dreadful British Airports Authority staff car park on the wrong side of LHR for me where I wait for an infrequent staff bus (that runs every fifteen minutes) then I have a ten minute journey into the LHR central area'.

## Temporary accommodation

Sixteen percent of respondents said they stayed in temporary accommodation when flying. Of these, fifty-seven percent stayed in a hotel, twelve and a half percent in a bed and breakfast, six percent in a rented flat, twelve and a half percent in a shared house, six percent in a second home and six percent in a motorhome. As to the quality of the accommodation, one quarter classified it as 'luxurious', one half as 'comfortable' and one quarter as 'basic'.

# Appendix 5

# Pilots' Commercial Flying Careers and Perceptions of the Industry

## Introduction

These data, derived from the questionnaire, concern pilots' commercial (as distinct from military) flying career and how pilots perceive or 'construct' themselves as individuals, and their industry. These constructions were adduced both directly (by asking: 'What personal rewards do you get from flying?' and: 'How do you see commercial aviation developing over the next five years?'), and tangentially (by asking pilots whether or not they would recommend a career in commercial aviation to their offspring).

## Commercial flying

Two percent of respondents had flown for very small-scale operations only. The remainder had flown for recognised air carriers (either air or helicopter lines). Of those respondents who had flown for recognised air carriers eighteen percent had had one flying job (either as a First Officer or Captain), thirty percent had had two flying jobs, sixteen percent had had three flying jobs, ten percent had had four flying jobs, four percent had had five flying jobs, nine percent had had six flying jobs, three percent had had seven flying jobs, five percent had had eight flying jobs, four percent had had nine flying jobs and one percent had had ten flying jobs. Put another way, at the time they submitted their questionnaires, roughly three-quarters of respondents had flown for between one and four air carriers.

Of those who spent two years or more at their first carrier (either fixed or rotary wing), twenty-five percent of respondents spent just two years, twelve percent spent three years, thirteen percent spent four years, twelve percent spent five years, two percent six years, four percent eight years and four percent nine years.

Of those respondents who worked for more than one carrier, twenty-two percent spent two years at their second carrier, seventeen percent three years, thirty-one percent four years, six percent five years, six percent six years, four percent seven years, four percent eight years and two percent nine years.

Of those respondents who worked for three carriers or more, twenty-one percent spent two years at their third carrier, twenty-four percent spent three years, fifteen percent spent four years, twelve percent spent five years and six percent spent seven years.

Of those respondents who worked for four carriers or more, twenty-seven percent spent two years at their fourth carrier, eighteen percent spent three years, five percent spent four years, twenty-three percent spent six years, five percent spent seven years and five percent spent nine years.

Of those respondents who worked for five carriers or more, fifteen percent spent two years at their fifth carrier, fifteen percent spent four years, fifteen percent spent five years, eight percent spent seven years and fifteen percent spent eight years.

The longest period spent by a respondent at a single carrier was thirty-eight years (actually their first carrier). Of those respondents who worked for more than one carrier, the longest period served in years for a second, third, fourth etc. carrier is given in Table 1.

**Table 1 Longest period served**

| Carrier | Longest period served (in years) |
|---|---|
| 2nd | 24 |
| 3rd | 17 |
| 4th | 21 |
| 5th | 17 |
| 6th | 17 |
| 7th | 8 |
| 8th | 11 |
| 9th | 6 |
| 10th | 4 |
| 11th | 5 |

Many respondents stayed for short periods (of between one and twelve months) at carriers. Forty-one percent of respondents stayed for between one and twelve months at their first carrier. Of those who worked for a second carrier, twenty-seven percent stayed for between one and twelve months. Of those who worked for third, fourth, fifth etc. carriers the percentage figures are given in Table 2.

**Table 2 Percentage who stayed for up to one year**

| Carrier | Percentage of pilots who stayed for up to 12 months |
|---|---|
| 3rd | 30 |
| 4th | 31 |
| 5th | 41 |
| 6th | 37 |
| 7th | 8 |
| 8th | 22 |

## Additional duties

There is a tradition in commercial aviation of involving pilots in non-flying roles. Thirty-three percent of respondents performed extra duties for their airline. These included, in respondents' own words, 'CRM training', 'quality assurance', 'maintaining navigational data', 'simulator training' and 'airline management'. Pilots were asked to describe the personal rewards they got from performing extra duties. A variety of responses were given, including: 'Pay!'; 'More money'; 'Extra salary: approximately £1,200 (I think)' [this respondent was a Flight Safety Officer for a fleet of Embraer regional jet aircraft]; 'NZD [New Zealand Dollars] 10,000 p.a.'; 'Financial of GB200 [assume pounds sterling] PCM [per calendar month]'; '£1,166 gross extra per month' [this pilot was a simulator instructor and examiner]; 'Slightly more pay (£1,500 per annum extra). Satisfying job'; 'Financial [see above] and sense of achievement. Satisfies need to lead'; 'Small financial and large job satisfaction rewards'; '18% extra pay. Satisfaction of improving standards' [this pilot did simulator training and worked for an Eire-registered freight carrier]; 'Satisfaction of helping others to improve their skills' [this pilot was involved in training]; 'I like training and have done so since RAF days'; 'I use my brain. It's a new company so I have to design new procedures. For example, I have just written a Training Captain course. I can be creative'; 'The satisfaction gained from imparting knowledge ...'; 'A sense of "doing good"; saving some errors; cash'; 'Financial. Know people. Visit places. Upgrade my knowledge and experience' [this pilot was Head of Flight Safety]; 'Hope of improving terms and conditions' [this pilot was a trades union representative]; 'Career improvement'; 'I get professional fulfilment from passing on knowledge and great satisfaction from successful remedial training'; 'I enjoy instructing but flight operations job is a drag' [this pilot was Flight Operations Manager]; 'Higher status. Better prepared for everyday flying ...' [this pilot had been a simulator instructor]. Two pregnant respondents restricted to administrative duties made the following statements: 'I can learn what happens "behind the scenes"'; 'More sociable hours and a lie-in!'. A 31-year-old pilot with a UK-registered 'no-frills' carrier who had no extra duties responded: 'I wish I could do some other duties!' A 44-year-old pilot with a Middle-Eastern long-haul carrier responded: 'None ... because I did it before and it provided very little reward for the effort, either financially or emotionally'. The former pilot had commenced his flying career in 1999 and had worked for two carriers. He had always flown out of the UK. The latter pilot had commenced his career in 1980 and had worked for three carriers. He had flown out of Britain, Malaysia, Indonesia, Trinidad, Nigeria and Dubai. A 39-year-old pilot flying out of Heathrow for a UK-registered carrier responded: 'None. I would like to [do other duties] but there are no opportunities .... I would like to make use of other skills – commercial flying four sectors a day out of LHR for five days each week is very wearing and repetitive. Not much scope for variety, creativity and the full use of mental capabilities'.

In response to the question 'In your opinion, are you allowed enough time to perform these other duties to *your* satisfaction?' fifty-nine percent answered 'yes' and forty-one percent 'no'. The following are comments from those who answered 'yes' but who appended conditions: 'Pretty much – all being well I was rostered one office day per fortnight'; 'Generally OK, but sim [simulator] rushed'; 'Consistent with commercial pressure – yes'; 'Yes, I have 10 office days per month (but work outside office hours)'. The following are comments from those who answered 'no': 'Never in a million years!' [this pilot was a Flight Operations Manager]; 'Absolutely no. Contracted and paid for 5 days/month, actually at least double required' [this pilot was a Flight Safety Officer]. Of those who answered 'no', the additional duties of ninety-two percent were directly linked with safe operation. These duties/jobs included 'maintaining navigational data for aircraft systems', Head of Flight Safety, Quality Assurance Officer, Flight Operations Manager, Crew Resource Management (CRM) Instructor, Simulator Instructor, Flight Safety Officer, Director of Operations, Chief Pilot, Training Captain. Thirty-six percent of those who performed extra duties had attended/were attending university. Fifty-one percent of those who did not perform extra duties had attended/were attending university.

## Perception of the commercial aviation industry

### *Rewards and directions*

Respondents were asked (a) 'What personal rewards do you get from flying?' and (b) 'How do you see commercial aviation developing over the next five years?' Here is a sample of the responses.

(a) 'I get a sense of achievement, I visit a variety of destinations and I get time at home with my family'. (b) 'Aviation will grow unless there are more major incidents (9-11, SARS [severe acute respiratory syndrome], etc)'.

(a) 'Camaraderie – a nice day out with friends. Technical satisfaction of a job well done'. (b) 'I foresee growing discontentment amongst those crews flying for regional airlines and the demise of a fairly major player in the UK/Europe'.

(a) 'I enjoy the actual flying. I enjoy the fact I have been able to cover usually 3,000 nautical miles in a twelve-hour day!' This pilot flew for GB Airways, whose destinations included Paphos and Marrakech. (b) 'I see increased competition. We will see more mergers between national carriers and small operators. Others will go out of business!'

(a) 'I absolutely love my job. Flying is not just a way to earn money, but a hobby/sport/pastime. I have been lucky enough to find a way to make a living. I never started flying to become an airline pilot. I just wanted to fly'. Asked about his worries this respondent mentioned 'excessive cost-reducing pressures leading to a curtailing of product quality and safety'. (b) 'I foresee consolidation of the majors, big layoffs, the rise of low-cost carriers and the decline in general of amenities'.

(a) 'I get financial security – although I was laid off after 9-11. People (still!) being amazed that a woman can fly a large jet aircraft'. This respondent commenced her working life as a bank clerk. She then worked as long-haul cabin crew for BA. (b) '[I see] less young people wanting to be pilots as they see the work getting harder and very tightly regulated ...'.

(a) 'I take pride in having a profession to maintain and in providing a service. Camaraderie'. This respondent mentioned fatigue as one of his concerns (he recommended that employers 'Listen to pilots!'). (b) 'I foresee more cost-consciousness, low-cost buy-outs and mergers. Hopefully rostering practices will stabilise to a point where companies are productive and pilots fit to fly!'. This respondent had flown for GO until it was bought by easyJet.

(a) 'Camaraderie; great people; enjoyment of flying; a sense of accomplishment'. This SkyWest pilot mentioned fatigue as one of his concerns: 'Scheduling often does not take into account issues of safety (i.e. a scheduled eight-leg, thirteen-hour day immediately following a reduced rest period with a time change)'. (b) 'Changes in the future will centre around the regional airlines. Success of major airlines will be a boon to the regional industry while failure or unrealistic demands of major partners may require regionals to sever partnerships, become independent and actually compete with their former partners. Proposed free-flight (in the USA) will allow unrestricted point-to-point service (mainly by LCCs and regional carriers) with very few delays. Long-haul (intercontinental) flying will still be carried out by major airlines from hubs that will still be plagued by delays. Lower-cost carriers will eventually find their way into this market as well, but possibly from satellite airports. "Dinosaur" carriers will not survive if they cannot operate efficiently or unless the industry becomes “re-regulated”’.

(a) ‘Previously I got camaraderie. Now it is just pay!’ (b) ‘Boom and bust. Boom and bust’. This ex-RAF Flight Lieutenant said he would recommend a flying career to his offspring.

(a) ‘Financial reward. Professional pride. A sense of achievement in a job well done. Team work. But I get very little recognition and minimal contact with passengers. Little praise from management, just constant hammering about our associated costs, and always the fear if it all goes wrong one day ...’. (b) ‘There will be continued attempts by managements to hammer pilots by reducing rewards and working even harder and further reducing the authority of aircraft crew members. They are trying to move commercial flying from being a “profession” to being more of a blue collar occupation’. This ex-British Telecom engineer had been flying for 14 years. He said he would not recommend flying as a career to his offspring. He emphasised the industry’s volatility: ‘I was made redundant by Air UK two months after joining in 1992, then I was re-hired two months later. I was made redundant by [my present airline] in January 2002 following 9/11, then taken on again’.

(a) ‘I get a sense of achievement (sometimes)!’ (b) ‘LCCs will price themselves out of the market. Fares will rise to realistic levels, subject to a possible terrorist attack’. This pilot flew for Thomas Cook.

(a) ‘Professional achievement’. (b) ‘More low-cost start-ups, saturation of ATC services, more slot delays before departure’.

(a) ‘I get financial security, a sense of achievement, camaraderie and a fantastic lifestyle’. This respondent flew for BA out of Heathrow. (b) ‘The A380 will herald a huge shift to larger aircraft, providing for less flights for the same number of customers. Also there will be a large increase in passenger numbers as memories of 9-11 begin to fade’.

(a)‘Financial security only’. (b) ‘Cyclic: general growth with occasional troughs’.

(a) ‘I love the physical handling of the aircraft, the power, speed, etc., I also enjoy the spectacular views. I also like other high-tech equipment: cars, computers, etc. A jet aircraft is just a big, expensive toy’. (b) ‘I think some LCCs will be taken over or go out of business. I also think many passengers will see the myth of “low-cost” and return to the full-service airlines (I hope)’.

(a) ‘I get a sense of achievement; I get to overcome challenges, meet new people, be with younger people [this respondent was 54 years old], be – more or less

– my own boss, be paid for something I love doing, be financially secure [this respondent had flown for BA from 1970 to 2004]'. (b) 'I foresee more delays due to increasing volumes of traffic'.

(a) 'I get a sense of achievement, camaraderie, financial security and I travel'. (b) 'The LCC sector will expand, the legacy carrier sector will decline. There will be an explosion of long-haul cargo, then a major slump. We will see the start of legislation to favour railways for short-haul beginning in the EU'. This respondent flew for GSS, an international cargo carrier.

(a) 'It's an enjoyable job – I like flying!!; I work with a pleasant bunch of people; It's better-paid than anything else that I am qualified to do'. This respondent was 49 years old. (b) 'The industry will become more polarised. The big "cushy job" airlines [the traditional legacy carriers] will become more entrenched and dictatorial while the bulk of the industry will be slogging it out ... and back-stabbing. Many new-starters will come and go'.

(a) 'Esprit-de-corps'. (b) 'Aircraft will be more automated; they will be bigger; they will, under normal circumstances, be simpler to fly'.

(a) 'It is an interesting, reasonably well-paid job that I can do. However, find me an equivalent job that does not involve unsocial hours and I will do it. Unfortunately for me the "magic" has gone out of flying'. This ex-RAF Squadron Leader had an Honours Degree in Computation. He had been doing a commercial flying job for four years. (b) 'More LCCs will start then fail'.

(a) 'I get camaraderie, security and a sense of having achieved something. It beats working for a living. We still get a fair amount of time off. For the moment I work 14–15 days each month'. (b) 'Ryanair and easyJet will continue to grow. We may see an increased presence of "cheap" labour from the East. It will not be easy to be a traditional airline with big overheads, expensive employees and powerful unions'. This respondent flew for SAS.

(a) 'I do a job that is usually straightforward, with a set task. I do a job that does not come home with me (except at sim-check time). I do a job that offers problem-solving challenges'. (b) 'I see more small operators going bust, and the amalgamation of the larger remaining players. I see longer hours with fewer perks for all airline employees – and not just in the low-cost sector'. This pilot, who had a Masters degree in Aeronautical Engineering (with First Class Honours), flew for easyJet.

(a) 'I get financial security, free travelling and camaraderie'. (b) 'There will be more cheap flights from the scheduled full-service operators. More airlines will go bankrupt'.

(a) 'I get financial security and friendship'. (b) 'Lots of LCCs will go bankrupt or be bought out. At least one charter operator will go bankrupt'. This respondent flew for First Choice.

(a) 'I get an enjoyment out of flying, financial reward and security'. This pilot flew for First Choice. (b) 'The holiday charter airlines will shrink and probably rationalise into smaller groups. The LCCs will compete with and drive each other into the ground and at least one will go bankrupt, probably after a major accident, given the training and recruitment trends, coupled with fatigue issues .... The majors will suffer the economic trends dictated by economic shifts and terrorism ...'.

(a) 'I enjoy flying. The job gives me reasonable financial reward and a comfortable lifestyle, with plenty of time at home (in comparison to the previous sixteen years in the RAF). I work with an excellent bunch of people in a job that gives variety'. This ex-RAF Squadron Leader flew for First Choice. (b) 'I foresee growth of the low-cost market, with a corresponding change towards niche marketing for the charter airlines. The majors will have to adjust their own business models in order to compete effectively with the LCCs, or move away from direct competition and concentrate on long-haul'.

(a) 'I get financial reward. I get a feeling of overcoming problems that self-serving management do not care to focus on'. This respondent had flown for Royal Brunei for over ten years. (b) 'I foresee bigger aircraft to alleviate traffic volumes. There will be less business travel (or reduced growth) due to electronic communication'.

(a) 'Pure flying enjoyment; the company of colleagues; financial benefit'. (b) I foresee public disenchantment with the low-cost model, and a return to a more stable environment with a solid economic base'. This respondent flew for Bmibaby.

(a) 'I never went reluctantly towards an aircraft. I enjoyed the work totally. I had a wonderful career. I did much committee work between 1988 and 1996 in the Guild of Air Pilots and Navigators and the Royal Aeronautical Society'. This 74-year-old ex-RAF Sergeant Pilot finished his career with Britannia (he flew for them for 18 years). (b) 'I foresee more and more leisure travel. Supersonics have probably had it now Concorde has been retired'.

(a) 'I get financial security, a sense of achievement ... plus I am fulfilling a childhood dream'. (b) 'I think we will see more mergers (like the one between

Air France and KLM) and a thinning out of all the low-cost airlines'.

(a) 'I get a sense of achievement and camaraderie'. (b) 'There will be more pressure for profits at the expense of safety'. This respondent flew for Air Mauritius.

(a) 'I have fun'. (b) 'The offshore helicopter industry will migrate to developing countries: the North Sea is on its way out'.

(a) 'I enjoy the beauty you see when breaking through cloud cover. Sunshine over the Alps in the early morning. Achieving a nice flight in total security'. (b) 'Majors will merge. Small companies will disappear. Secondary airports will have fewer flights'.

(a) 'I enjoy the challenge and sense of achievement'. (b) 'There will be reduced use of business travel and more leisure travel. The public use aircraft now like they would a bus. They no longer see it as a glamorous or special way to travel, and that perspective will continue'.

(a) 'I get a sense of worth and achievement'. (b) 'There will be further expansion of LCCs into Eastern Europe, and possibly trans-Atlantic arenas, alongside a further expansion of private/corporate business jet operators'.

(a) 'I get a good salary, good time-off, I'm very seldom away from home, my work fits with family life'. This respondent worked for CHC Scotia Helicopters. (b) 'There will be a gradual increase in demand for pilots, balanced by pilots working to an increased age before retirement'.

(a) 'I get financial security, camaraderie, a variation in lifestyle ... It's good fun!' This pilot flew for Air Contractors out of Dublin. (b) 'I see an expansion of routes, more LCCs, lower fares but poorer quality'.

(a) 'My satisfaction is knowing I'm doing the best job in the world'. (b) 'I foresee low-cost long-haul. Several LCCs will disappear. European national carriers will merge'.

(a) 'I get camaraderie, a sense of achievement, financial security and social respect: In Asia piloting is still seen to be a good job'. This respondent had been flying for Cathay Pacific for 24 years. (b) 'I foresee greater automation (crews will not be allowed to fly manually); there will be more accidents with third-level carriers; there will be a broader spectrum of choice in quality and safety. Commercial aviation will come to resemble the hotel market. Some will be cheap and nasty, some will be five star. There will be clients for all.

There will be long-haul low-cost carriers (e.g. UK to Australia)'.

(a) 'I get financial and professional rewards'. (b) 'More aircraft, cheaper pilots, more accidents'. This respondent flew for First Choice.

(a) 'My rewards are man–machine interaction and camaraderie'. (b) 'I foresee no major changes'.

(a) 'We all have to pay the mortgage – it beats working for a living'. This pilot flew for Britannia. (b) 'Low-cost airlines and charter operators will find a balance. More pilots will work part-time – getting sick of ever more complicated operations and security rules, etc. United Kingdom passenger traffic will stay at about the same level, despite forecast growth'.

(a) 'It used to be camaraderie, a sense of achievement, etc., but now it is none of them. I am looking for a career change. Pilots and beancounters [accountants] do not mix and the beancounters have taken over the industry'. This ex-RAF Flight Lieutenant flew for Royal Brunei. He commenced his commercial flying career with First Choice in 1994. (b) 'I foresee more cost-cutting, more pension schemes scrapped, more accidents, more rich Chief Executive Officers and more poor pilots'.

(a) 'I was never able to work indoors all day. When flying my office scenery changes, and I feel a greater sense of accomplishment than I would punching a clock eight hours a day'. (b) 'I think there will be a lot of consolidation within the industry with carriers being bought up or merged. Legacy airlines (like United, American and Delta) will either be forced to compete with the LCCs by totally restructuring their business models and abandoning the hub-and-spoke system, or they will have to abandon much of their domestic flying and leave it to the LCCs while they concentrate on hub-to-hub and international flying. Regionals will either become LCCs to provide somewhat more stability by distancing themselves from bankrupt or potentially bankrupt partners, or be folded in to those partners as part of a consolidation'.

(a) 'I get the satisfaction of a job well done due to the appreciation shown to me by my superiors. My financial rewards are commensurate with my level of aircraft'. This pilot flew for SkyWest. (b) 'I foresee the re-regulation of the industry, or the demise of several carriers, along with industry consolidation'.

(a) 'It gives me financial security and my schedule allows for a flexible lifestyle'. This respondent flew for Icelandair. (b) 'Low-cost might be good for the travelling public, but at what cost? Corners will be cut in maintenance and crew training (pilots might be required to pay for their own initial and recurrent training). Crews will be required to put in more hours leading to fatigue. All this and more is a way to increase profits for shareholders'.

(a) 'I get a sense of achievement and self-improvement ... I get a sense of an ever-expanding environment requiring continuous development'. This respondent flew for Emirates. (b) 'I foresee bigger aircraft, longer routes, synthetic external views, more computerised flight-path systems – both ground and air-based. Piloting will become more and more a monitoring function'.

(a) 'It's a job that I have always wanted to do. I love to fly and that has not changed. The salary is good and people are friendly. I enjoy being part of a team'. This respondent, who had an upper second class Honours degree in geography from Kings College in London worked for Bmibaby. She said she would recommend a career in commercial aviation to her children. (b) 'There will be more super-carriers and fewer small operators and start-ups. There will be an increasing demand for cheap flights, long-haul included. Airspace congestion will increase'.

(a) 'I only do ad-hoc flying now ... I get satisfaction that I can still do it'. This 64-year-old ex-RAF Flight Lieutenant was flying ad-hoc duties for Air Contractors out of Dublin. (b) 'I foresee a reduction in "no-frills" services, and an improved quality of service from BA-type airlines. I expect a major no-frills failure'.

(a) 'Sense of responsibility'. (b) 'I foresee corporate instability, i.e. bankruptcies. There will be consolidation amongst the LCCs. There will be at least one major air disaster in the UK with a human factors cause'.

(a) 'There is some adventure, still, and money in the bank'. This 55-year-old pilot, born in Sweden, had lived in Cyprus, Germany and Sweden. (b) 'In the next five years we will have more corporate grouping, and a kind of monopoly will arise again'.

(a) 'I get camaraderie, monetary reward, etc'. (b) 'I foresee increasing congestion and relentless commercial pressure combining to erode safety levels. There will be a continued erosion of Flight Time Limitations'.

(a) 'I get camaraderie, monetary reward, a sense of achievement, a purpose in life, and a social outing. Every landing at LBA [Leeds-Bradford International Airport] is a challenge. Self-worth'. This respondent flew for Channel Express. (b) 'I foresee the amalgamation of airlines, survival of the fittest and a deterioration in pay and conditions'.

(a) 'I get financial security, a vast sense of achievement and personal well-being, camaraderie, enjoyment of the task itself, and I get to travel to interesting places'. This respondent flew for SkyWest. (b) 'I foresee LCCs supplanting the older, mainline carriers. There will be a transition to smaller aircraft and regional jets for domestic services. This will leave international services to a few, very large aircraft. In five years there will be only two to four US domestic carriers'.

(a) 'I get instant satisfaction after completing a flight, financial security, social status and fun. I am constantly challenged by my younger colleagues'. This 46-year-old pilot had flown for Austrian since 1978. (b) 'I foresee fragmentation on the routes being flown; a gradual move by major carriers to copy the LCCs; the introduction of ultra-long range sectors; downsliding yields'.

(a) 'I get camaraderie, reward, etc'. (b) 'I foresee an increased number of incidents/accidents reflecting commercial pressure on rostering and experience levels on the flight-deck'.

(a) 'I do not get any personal rewards any more. It is just a job that I am glad to go home and forget about!' This 45-year-old ex-sales representative started his commercial flying career in 1983. In 2004 he was flying for Monarch Airlines. (b) 'The low-cost bubble may burst'.

(a) 'I get financial security, a challenge, camaraderie and a fairly good lifestyle that affords me time with my family'. This respondent flew for Dragonair. (b) 'I think LCCs will thrive. Several major airlines will go bankrupt. Longer-term I see pressure mounting on LCCs as the pool of available pilots dries up and higher competitive salaries and benefits raises the LCC cost structure'.

(a) 'I enjoy the actual flying, the sensation, the view! I enjoy the achievement of doing a complex job well'. This respondent, who had a PhD in microbiology, flew for BMI Regional. (b) 'I foresee the continuing rise of LCCs leading to further pressure on inclusive tour and regional airlines'.

(a) 'I get financial reward'. (b) 'There will be even more profit-orientation and exploitation of the workforce'.

(a) 'I started flying in 1954 at school with an RAF flying scholarship, and from then on have never lost the taste!' (b) 'I foresee the growth of low-cost; fuel-meagre aircraft; no more controlled flight into terrain (CFIT); free-flight with global positioning systems and traffic alert and collision-avoidance systems releasing ATC to perform a more general overview'. This ex-RAF Flight Lieutenant retired from easyJet in 2001 at 65 years old.

(a) 'I get enjoyment and satisfaction'. (b) 'There will be fewer companies, fewer jobs and more regulation. There will be a continuing reduction in piloting skills'. This ex-RAF Flight Lieutenant had flown for Manx Airlines for over two decades.

(a) 'I get paid; I like flying; I have an office with an amazing view; I am living the boy's dream'. (b) 'More LCCs will come and go. I do not think the last major company has gone belly-up. However the bigger alliances will help the majors to gain control over the market. In five years I believe that aviation will be run by both airline managements and the courts – telling what is allowed and what is not. I believe the industry will slow down because of this'. This respondent flew for Ryanair out of Charleroi (Brussels South).

(a) 'I have a great job that gives me satisfaction. I get time with my family. I get financial benefits. Our travel scheme is good [Dragonair]'. (b) 'There will be: increased traffic; a pilot shortage; larger aircraft; an erosion of safety margins; reduced pilot-experience levels'. This respondent, who had a degree in economics, flew for Dragonair out of Hong Kong.

(a) 'I get professional satisfaction from a flown approach (or similar). I get to visit new places. Variety'. (b) 'There will be bigger long-haul aircraft; more competition driving costs (pay and conditions) down; more short-term contracts; more pilots from new EU countries prepared to work for virtually nothing'. This ex-RAF Flight Lieutenant, who had a degree in mathematics, flew for First Choice.

(a) 'Despite the fact that I still like flying, all the really good motivations and enthusiasm are long-gone – with the exception of two. All my employers have demonstrated how badly a pilot can be considered and treated, especially in countries where social laws and work-ethics and regulation are weak'. This ex-Armee de l'Air pilot flew for Air Mauritius. (b) 'Traffic will rise. There will be (again) a shortage of qualified and experienced pilots. The age limit will be pushed to 63. Apart from major carriers where pilots may be able to negotiate decent conditions and salaries, generally there will be fewer opportunities for long-term career planning. Corporate flying will become the best type of flying'.

(a) 'I get enormous satisfaction from using my skills in a complex and demanding task. I get a great sense of social involvement and pride working amongst folks from all walks of life at the airport. I get financial independence for my family'. (b) 'I foresee more pressure on pilots from a lack of them, and a lack of revenue to pay them well. There will be dire security problems posed by man-portable surface-to-air missiles which could bring air travel to an end for a long period'.

(a) 'My rewards are purely financial'. (b) 'There will be hopefully larger aircraft thereby cutting the volume of flights'.

(a) 'Flying is reasonably well-paid. I get to mix with like-minded people. I get a very occasional sense of achievement!' This ex-Royal Navy Fleet Air Arm Flight Lieutenant who had a degree in civil engineering had been flying for First Choice for over a decade. (b) '[Flying will become an] increasingly volatile career; less security; less reward'.

(a) 'I get financial reward and enjoyment from flying aircraft'. (b) 'There will be more congestion, at airports and on motorway and railway-links, and in the air. Salaries will continue to erode, as will working terms and conditions. Fewer people will be attracted to the career when they compare the better rewards and lifestyles offered by other professions. Airlines will continue to cut costs to meet public expectations .... The implications of this are obvious, and most likely detrimental to the industry'.

(a) 'I get an OK salary; satisfaction if flying under difficult conditions; a good working atmosphere'. (b) 'I foresee pilots losing benefits and more job insecurity; pressures on crews will increase due to competition and pressure to survive; lower standards of service for passengers due to lower ticket prices; but full-service will make a come-back for passengers willing to pay; more airline failures; reduction of pilot-status and erosion of respect for what pilots do'.

(a) 'I get a sense of achievement and enjoyment and financial security'. (b) 'There will be more LCCs and more flexibility for consumers. Pilots will be working harder. They will be doing longer hours'.

(a) 'Flying is the only thing I have ever wanted to do. I love it'. (b) 'There will be more automation, bigger aircraft and longer hours'.

*A future career for my offspring?*

Thirty-five percent of respondents said they would encourage their offspring to become a commercial pilot. Sixty-five percent said they would not. Of those respondents born in the UK, thirty-six percent said they would encourage their offspring to become a commercial pilot. Of those respondents born outside the UK, thirty-two percent said they would encourage their offspring to become a commercial pilot.

This is a representative sample of comments from respondents who said they would encourage their offspring to become a commercial pilot. Where respondents cited good living standards as a reason for recommending a flying career, their financial circumstances (specifically the value of property and number of cars owned) are provided. (Valuations are given only for properties owned outright by the respondent.)

'Although people grumble and the job is changing it is still a sound profession and provides a reasonable income.' This respondent had a live-in partner who worked as a pilot. They owned a three-bedroomed house in England valued in 2004 at £300,000. The respondent owned a VW Golf TDI 1.9. There were no children.

'It is ... a great alternative to the regular rat race ...'.

'Best office in the world ...'.

'Reasonably well paid, extremely varied [this pilot flew for British Airways out of London Heathrow Airport (LHR)], meet lots of people, keeps you mentally active'. The respondent and his homemaker-wife lived in England in a three-bedroomed house valued in 2004 at £250,000. They owned two cars, a Ford Cougar 2.5 V6 and a Fiat Marea 1.8. They had two infant children.

'It is an honest career. I make adequate (not great) money, even at the regional level'. This respondent had three children, all at school. He and his working wife (a computer programmer) lived in a four-bedroomed house in the USA valued in 2004 at $180,000. They also owned a five berth boat valued at $40,000. They owned three cars, a Ford Mustang, a Chevrolet Suburban and a Toyota Corolla.

'Yes. Financial security [this pilot flew for GB Airways, a regional and international scheduled carrier] '. He lived in England in a three-bedroomed rented house and drove a VW Golf GT TDi.

'Yes. Allows for great personal growth'.

'Best job in the world'

'Safe, well-paid job with excellent lifestyle potential, excellent possibilities to develop your career away from the flight deck [this pilot flew for British Airways out of LHR] and a job that still appears to have at least some respect in the outside world'. The respondent and his wife (who worked as an Operations Manager) lived in England in a four-bedroomed house valued in 2004 at £330,000. They part-owned a second or holiday home in England (a three-bedroomed house) valued in 2004 at £160,000. They owned two cars, a Volvo 2.3 GL Estate and a Honda Civic 1.6 SE.

'Why not? It is still a worthy profession'.

'It's the best fun to be had and be paid for'.

'Relatively well paid. Not short-term repetitive. Great views'. This respondent and his wife lived in England in a three-bedroomed house valued in 2004 at £180,000. They also owned a three-bedroomed house in Wales valued in 2004 at £200,000 and had a one-third share in a house in France. They had no children.

'Yes. I came through the best times [this pilot started flying in 1971 with British Airways at LHR] but there are still good opportunities for those who love flying'.

'Yes. It's a passion'.

'Yes. Good lifestyle, if long hours. Huge job satisfaction. Talented work force'. This pilot flew for Excel Airways, an international scheduled and charter carrier.

'Yes. It is an interesting and financially rewarding career'. This 24-year-old respondent, who did not have a partner, lived in England in a three-bedroomed house valued in 2004 at £180,000. He drove a Peugeot 206 Hdi. He flew for Britannia Airways out of Manchester, Nottingham East Midlands and London Gatwick.

The following is a representative sample of comments from respondents whose responses were ambiguous or whose recommendation was conditional.

'Maybe. If they showed real motivation I would encourage them'.

'Probably wouldn't encourage it; too expensive to get in to; no job security; dependent upon passing a medical every six months ... better to just get a PPL [Private Pilot's License] and do it for fun'. This pilot flew for Channel Express, a UK-registered domestic, regional and international charter operator.

'Yes ... provided that they set up their own business beforehand'. This pilot appeared also to run a business.

'Yes. But only if they show an interest. It's a fairly good job for a working-class person'. This 40-year-old pilot, the son of a greengrocer, left school at 16. His jobs included apprentice toolmaker, apprentice mechanic and labourer.

'Yes ... because I had a wonderful career. But my daughter once said: "What, like you, Dad, a waitress living out of a suitcase" when I suggested she try being a stewardess!!'

'Yes ... but only if they start early and follow a career in a major [airline]'.

'If they want to I would support them'. This respondent went on to say that he would recommend a military flying career because he had enjoyed his military flying (he had flown for the Royal Air Force). He had two sons, one ten years old the other seven.

'If they want to become a pilot I would support them, but I would not encourage them. The industry is becoming more and more ruthless'. This pilot lived in a four-bedroomed rented house with his live-in homemaker-partner. The couple had a two-year-old son. He flew for Ryanair.

'I would encourage them to fly if they want to, though only to be commercial pilots as a means to stay flying. It is a vicious industry and I have been lucky, though I know many who have not done as well. Those who make a good living out of flying are in the extreme minority'. This respondent, who had a degree in philosophy from Berkeley, flew for SkyWest Airlines, a US-registered scheduled regional and domestic carrier. Recommended alternative careers were medicine, architecture and real-estate development. According to the respondent these careers provided 'The opportunity to have an intellectually rewarding job which provides income to pursue hobbies ... '. The respondent lived with a working partner in Fresno, California in a rented two-bedroomed apartment for which, in 2004, they paid $700 per month. They owned a Chevrolet S-10. They had no children.

'Yes ... still the best job in the world, once you accept that most of the owners/management really suck [sic]!'

The following is a representative sample of comments and suggested alternative careers from respondents who said they would not encourage their offspring to become a commercial pilot. Where respondents cited falling living standards as a reason for not recommending a flying career their financial circumstances (specifically the value of property and number of cars owned) are provided. (Valuations are given only for properties owned outright by the respondent.)

'Career rewards seem to be diminishing in both lifestyle and financial reward'. This father of two, who had an MBA, recommended accounting and the law. According to the respondent these careers offered 'reward and lifestyle, though I would recommend pursuing a career which they enjoyed for reasons other than just financial'. The respondent had two young sons. He lived in Scotland in a four-bedroomed house valued in 2004 at £700,000. His previous dwelling, a three-bedroomed flat, was valued at £400,000. He and his homemaker-wife had a second or holiday home (a two-bedroomed flat) in Scotland valued at £250,000. They owned two cars, a VW Golf VR6 and a Jeep Grand Cherokee. He flew for First Choice Airways out of Glasgow.

'Rewards do not justify the financial input required'. This father of three, who had an Ordinary National Certificate (ONC), preferred his children to go in to the professions or, as he put it, 'anything they are interested in'. He and his homemaker-wife lived in England in a six-bedroomed house valued in 2004 at £1,000,000. (His previous abode, a four-bedroomed house, had been valued at £350,000.) They owned two cars, an Audi TT and a Land Rover Discovery.

'Piloting is a very insecure career as a fail by the operator or of the medical can have drastic results for the individual pilot who normally has no other skill or qualification'. This 76-year-old respondent who started commercial flying in 1957 after leaving the British Army, had two sons and a daughter. The eldest son was a Finance Director, the youngest was a senior manager for IBM Europe. His daughter was a homemaker. Both his sons had attended university. His daughter had not. This respondent's recommended careers were Medicine, Law and Banking/Finance. In answer to the question 'Why would you recommend these careers over others?' he responded: 'Any of the above offer security during the career and also a large variety of different skills can be employed. Pilots only have one basic skill ... and when they cannot use that they are lost. There are no limitations on grounds of health or age, both of which of course are severe for pilots'.

'It's not flying any more: it's managing a computer and hoping that you never actually have to use your flying skills if something goes wrong'. This respondent, who had an

Honours degree in Computer Science from Manchester University and a teenage son and daughter, recommended engineering and medicine.

'It is a bad career for health and family reasons'. This respondent recommended 'Any of the other professions' because of their 'Regular lifestyle and stability'.

'It's not what I believed it to be, as the profession has been changed by the actions of airlines and inaction of the Civil Aviation Authority over the last twenty years or so. [It is] not sufficiently broad and challenging for a bright graduate candidate. Uses too narrow a range of skills in too repetitive an environment. High-workload environments with little time off and very anti-social shift-work are not good for individual health and family. The profession is not rewarded in time off or remuneration or recognition as it used to be. It has fallen behind Doctors, Lawyers, etc. in status and reward'. This respondent recommended medicine, science and engineering (he had a BSc in Electronic Engineering) or the law because these professions were 'More intellectually stretching and challenging. More job satisfaction. [Gives] a proper identity to the individual – not just a number on the seniority list out of hundreds or thousands. [There is] interaction with society and the public. Commercial flying is quite a lonely and isolated affair, particularly after 9/11'.

'Job uncertainty, big initial overhead [the cost of training], difficult to progress'. This respondent, who had a BSc degree, recommended medicine, dentistry and the law. His first son was a storeman and his second an electronics technician.

'Being a commercial pilot is not as lucrative as other professions, bankers, doctors, etc'. This respondent, who left school without taking A-levels, recommended banking, the law and medicine, on grounds that 'In these occupations reward is commensurate with study and responsibility'. The respondent, his working wife and their two young children lived in England in a three-bedroomed house valued in 2004 at £360,000. They owned a second home (a two-bedroomed house) from which they derived rental income. This was valued in 2004 at £320,000. They owned three cars, a 1997 VW Passat Estate, a 1990 VW Passat Estate and a 1973 VW Beetle 1303S Cabriolet.

'They do not have what it takes'. This respondent had a teenage son and daughter. He recommended any skilled manual job because 'There will always be a demand for those who can do things with their hands'.

'Continual drive by managers to cut costs, health issues with high altitudes, loss of status/responsibility'.

'It's not the same career it used to be. Pilots are required to be away from home more and the pay is not what it used to be'. This respondent recommended the law or medicine as alternatives 'Because then my children could afford to own a plane and fly for pleasure'. The respondent and his working wife lived in the USA in a six-bedroomed house valued in 2004 at $230,000. They owned two cars, a 2003 Mazda Protégé and 2001 Isuzu Rodeo. They had no children as at 2004. The respondent flew for SkyWest out of Denver.

'Unless you can get in to a "major" ... the poor job security and lifestyle is not worth the investment'. This respondent recommended teaching, the law and 'professional engineering or science research' on grounds of 'lifestyle and job stability'. He flew for

Eastern Airways, a domestic and international scheduled and charter carrier, out of Leeds/Bradford Airport. During his 25-year flying career he had been made redundant twice, had resigned three times and had been on a fixed-term contract four times.

'The future status of airline pilots is being destroyed by Airbus'.

'Increasingly volatile career; less security; less reward'. This respondent, who had two daughters, recommended journalism, general media or the law as alternative careers. In his view they offered 'A rich variety of work; you are always in demand; you have the chance to try different areas within these careers'. He and his homemaker-wife lived in England in a four-bedroomed house valued in 2004 at £450,000. They owned a Lexus 200 SE. The respondent flew for First Choice Airways.

'Poor financial rewards compared to other professions; anti-social working hours; work is less stimulating due to high levels of automation; detrimental effects on health due to poor working conditions (i.e. poor diet); inability to relax due to time pressures'. This respondent recommended accountancy, the law or 'entrepreneurship and philanthropy' as alternative careers. His justification was that they offered 'Better lifestyles; not continually assessed once qualified; better financial rewards; better time off once a more senior position is attained – unlike a senior airline Captain'. This respondent came from an upper-middle class background. His father had been a civil engineer and his mother a radiographer (later a Superintendent Radiographer). He had attained a BEng (Hons.) in Aeronautical Engineering from the University of Glasgow. He flew for BMI out of Heathrow. He lived alone in a three-bedroomed flat valued in 2004 at £250,000. He owned a Rover MGF.

'My children [a boy and girl] are unlikely to have a secure future in view of the proliferation of LCCs'.

'My children are too studious and academic to really fit in as pilots'. This respondent had four daughters. All had attended university. Two were doctors, one was an accountant and one a hotelier (with her husband). The respondent himself had left school after A-levels.

'Remuneration is decreasing. It is very competitive. Very expensive to train. Unhealthy environment'. This respondent recommended 'Whatever makes them happy'. He had three daughters. His eldest daughter was a journalist. His middle daughter was at university. His youngest daughter was at home. He and his homemaker-wife and their daughter lived in England in a four-bedroomed house valued in 2004 at £320,000. They owned two vehicles, a Peugeot 206 HDi and a Fiat Ducato motorhome.

'It is no longer the profession I joined nearly fifty years ago'. This respondent had started flying with British European Airways (BEA) in 1963. He had been a Flight Lieutenant in the Royal Air Force. He had two sons and two daughters. Two were gem dealers, one was a computer analyst and one a homemaker. None had attended university. The respondent had left school after A-levels.

'The job of pilot is losing its appeal with fewer jobs available after very expensive training'. This respondent recommended the financial sector, medicine and information technology on grounds that they might offer a more stable career.

'The work is getting harder and very tightly regulated due to security constraints, etc'. This respondent said he would encourage his offspring to go into medicine, dentistry or information technology. She described IT as 'today's career'. She said that doctors and dentists 'were always needed'.

'The industry only cares about money, not people. Wages are dropping'. This respondent had one son. He was studying maritime engineering at university. The respondent and his working partner lived in a three-bedroomed house (in Norway) valued in 2004 at £50,000. They owned a second or holiday home in Norway (a two-bedroomed house) also valued at £50,000. They owned three cars, a 2003 BMW 320 Touring, a 2002 Nissan X-Trail and a 1992 Saab 900.

'Conditions are getting worse. The good times are over ... I hope low-cost carriers don't find pilots willing to "fly for food"'. This single 38-year-old respondent started her career as a flight attendant. Her permanent residence was a two-bedroomed flat in Berlin valued in 2004 at $100,000. She owned two cars, a 1999 Mercedes E300 and a 1997 Nissan Micra.

'Financial rewards no longer suitable for a comfortable life-style'. This pilot, who had a BSc, flew for Thomas Cook Airlines (UK) out of London Gatwick Airport. The Respondent's permanent residence was a two-bedroomed flat valued in 2004 at £275,000. The respondent's non-cohabiting partner worked as a surveyor. The respondent owned a VW Polo.

'The cost of education today is much higher, the starting pay is much lower, we work longer hours and future retirement plans will be much worse'. This respondent recommended the medical profession and fashion industry. He had a young son and daughter.

'It's becoming a difficult job, with ever-declining rewards and decreasing safety margins. It is very unhealthy also'. This respondent recommended writing, teaching or acupuncture on grounds that such jobs involved helping people and being creative. Also they provided 'more control over your life'. The respondent had a one-year-old daughter. He and his homemaker-partner lived in Eire in a two-bedroomed house valued in 2004 at $350,000. They owned a Lancia Dedra 2.0 ie.

'The pressures are greater now. Limits – regarding flight-time limitations – have become the normal working schedule. The pay is now not so good, especially considering the hours worked. My working hours, when considering the time available to me at home, equated to a sixty-hour week. I would not wish to subject my children to a lifetime of that. In my opinion airlines like Ryanair are an accident waiting to happen – just review the PPrune web site [PPrune (Professional Pilots' Rumour Network) is a web-based discussion forum] – and flying for these airlines is, quite often, the only route to building the hours necessary to obtain a decent flying job'. This respondent had two daughters and a son. Both daughters worked in public relations. The son was a sales assistant. Both daughters had been to university. The pilot and his working wife lived in England in a five-bedroomed house valued in 2004 at £700,000. They had a four-bedroomed second or holiday home in the Turks and Caicos Islands valued in 2004 at $395,000. They owned three cars, an Audi A4 Avant, an Audi 80 and a Peugeot 306.

'I would never try to sway them into this profession. It's uncertain and tough on families .... Life isn't only about what you do for a living, it's about searching for and finding happiness. I think too often people forget about the big picture, about what a career can mean to their family/personal lives'. This respondent, a graduate of Embry Riddle aeronautical university, flew for SkyWest, a US scheduled domestic and regional carrier.

'The lifestyle and stress of the job are not suitable for my two children'. This respondent, a graduate of University College Cardiff in Wales (where he achieved a BSc (Hons) in Mechanical Engineering), flew for First Choice Airways, a UK-based international scheduled and charter carrier. He had two daughters, one fourteen the other ten. He had flown for the Royal Navy, leaving with the rank of Lieutenant Commander.

'Lack of job security and an uncertain future, especially with the demise of fossil fuels'. This respondent had a BSc (Hons) in Mechanical Engineering from Edinburgh University in Scotland. He had two sons, one twelve the other ten. He recommended the professions, specifically law and accountancy, for two reasons. First, 'Because the majority of major business players have this background'. Secondly, 'Because happiness in the job is a major factor when considering a career'. He flew for First Choice Airways.

'The job is becoming too insecure. Huge cost of getting your licence with no guarantee of a job. World's oil is running out'. This respondent, who was divorced, had two daughters, one ten the other seven. He flew for Monarch, a UK-based scheduled and charter carrier. The respondent claimed that his marriage was affected by 'Constant weekend work [and] multiple roster changes'.

'Lack of respect for the job by airline management; stress; shift patterns; job instability; insecurity caused by low-cost airlines'. This respondent's father had been an airline pilot. The respondent had spent one year at university before starting an Air Transport Pilot's Licence (ATPL) course. He flew for Britannia Airways, a regional and international charter carrier, out of Birmingham, Manchester and Nottingham East Midlands Airport. He had no offspring.

'I believe they can do better for themselves outside of aviation'. He had three daughters. Two worked in Information Technology Consulting. The youngest was still at university. The respondent flew for Nippon Cargo Airlines.

'NO NEVER [sic]. Appalling lifestyle'. This respondent, who had two young daughters, flew for Monarch Airlines. He elaborated: 'I'm paid £100,000 per year but I definitely do not feel valued as an employee. My boss ignores me so there's no communication. I feel little loyalty to my employer'. He came from a working-class background. His father had been a toolmaker. His recommended careers were law, journalism and 'self-employment – plumbing, carpentry, tiling, etc'.

'The rewards are not any more up to the efforts (studies and personal life) and amounts of money involved'. This respondent had a son and daughter. His son was in the French Army, his daughter was a student in Rennes, France. The respondent had commenced his flying career with the French Air Force where he became a 'fighter combat formation leader'. His recommended careers were 'anything in the human sciences', architecture, civil engineering or green farming. In his opinion these jobs 'could contribute to the progress of mankind'. He flew for Air Mauritius, an international, regional and domestic scheduled and charter carrier. Although based in Mauritius, he owned a four-bedroomed house in France valued in 2004 at $350,000.

'They are all girls! There is no other job I would ever like to do but flying [but] if you don't have that passion for the job you are left with insecurity, mediocre pay, shift work, stress, anxiety, etc'. Regarding his recommended alternative careers he stated: 'I don't have a specific recommendation but would advise them to opt for a job they enjoy rather than high income, as you spend a large chunk [portion] of your life at work'. This pilot flew for BMI Regional, a UK-registered international, regional and domestic scheduled and charter carrier. Before becoming a pilot he had been a trainee quantity surveyor, a milkman [someone who delivers fresh milk door-to-door] and a production operator for the Ford Motor Company. He had three daughters. One was a part-time waitress. The other two attended secondary school. He and his working-wife (a nurse) lived in a four-bedroomed house in England valued in 2004 at £320,000. They had two cars, a Rover 75 and a Fiat Punto.

'Too much "job" and not enough "vocation". The training costs for self-sponsorship are excessively burdensome and salaries are falling. Disrupted lifestyle and a high stress environment'. This respondent, who had flown for the RAF, had a mature son and daughter. The former was a university student, the latter self-employed. He worked as a freelance pilot at Biggin Hill Aerodrome in the UK. He and his homemaker-wife lived in a four-bedroomed house valued in 2004 at £375,000. They ran a Toyota Corolla.

'The profession at the moment is under constant threat of cost-cutting .... Additionally the value of being a pilot has diminished whilst the responsibility has not'. This respondent, whose father had been an airline pilot, recommended the medical profession, self-employment or 'anything my children truly desire'. He had twin sons aged fifteen months.

'Long training and development for little reward. Poor career progression. Hard lifestyle (unsocial hours). Incredibly tedious routinely'. This respondent, who came from a working-class background (his father had been a master plasterer and his mother a nurse) gained a degree then went on to attain the rank of Squadron Leader in the Royal Air Force. He had a teenage son and daughter. His recommended alternative occupations were the uniformed services (for example, the police service or military), medicine or tourism and leisure. He believed these careers to be more stable, to offer more variety and better prospects. He flew for First Choice Airways out of Glasgow. He and his working wife (a personal assistant (PA)) lived in a four-bedroomed house valued in 2004 at £300,000. They owned two cars, a Renault Scenic and a Fiat Punto.

'No because I would have to pay for it. There is no money in it anymore'. This respondent recommended 'owning your own business' on grounds that such a career offered 'financial independence and stability', albeit through 'hard work'. He and his working wife (an aircraft dispatcher) lived in England in a three-bedroomed house valued in 2004 at £200,000. They owned two cars, a VW Golf and a Ford Fiesta. They had no children.

'Work is too fatiguing for the rewards on offer'. This respondent recommended the law and engineering as alternative careers because they offered 'Good pay and a good deal of respect for the skills'. He had a teenage son at school and a daughter at university. He and his working wife (a swimming teacher) lived in a four-bedroomed house valued in 2004 at £330,000. They owned two cars, a Toyota Corolla and a Ford Mondeo. The pilot flew for Monarch Airlines.

'Less young people wanting to be pilots as they see the work getting harder and very tightly regulated ... whilst salaries and packages seem to be getting cut and streamlined, e.g. final pension schemes disappearing, etc'. This respondent recommended medicine, dentistry and information technology as alternative careers. She and her working husband (a pilot) lived in England in a five-bedroomed house valued in 2004 at £600,000. They had two second or holiday homes, one a three-bedroomed house valued at £150,000 the other a two-bedroomed flat valued at £150,000. Both were located in England. They owned two cars, a BMW III-Series Coupe and a VW Golf Cabriolet 1800. They had no children.

'The days of innocent aviating are numbered. We now manage the systems. It is too regulated and automated for absolute job satisfaction. Plus the pay is not superb when you consider the on-going evaluations and medicals'. This respondent recommended environmental science, telecommunications and the media on grounds of 'satisfaction, security and wealth creation'. He had a nine-month-old daughter. He lived with his homemaker-wife in a four-bedroomed rented house in Mauritius. They ran a Mitsubishi 4wd.

'Commercial flying is a physically difficult job with little chance of success'. This respondent, who came from a working-class background, recommended medicine, the financial sector or the air force on grounds of 'job security and good prospects'. The First Officer owned a two-bedroomed house valued in 2004 at £200,000 and had a non-cohabiting partner who worked as a Financial Director. The respondent flew for DHL Air (UK).

'The job has become a grind. To keep up the pace of work required by today's operators for a whole career would be health-threatening'. This respondent recommended government or the military as alternative careers on grounds that government service 'had fewer commercial pressures, better conditions and good pensions'. This pilot flew for Bmibaby, a UK-registered low-cost carrier.

'No. It would be financially ruinous for them. Although there is enormous satisfaction to be obtained from being a pilot, I often get to think the clever folks are the ones who can afford to sit down the back and pay for it'. This respondent recommended finance or self employment because, as he put it: 'The world seems to value people who control companies and money far more than it seems to value those of us who merely control aircraft in flight'. The respondent and his homemaker-wife lived in England in a five-bedroomed house valued in 2004 at £300,000 (in 1983 the property had been valued at £60,000). They owned two cars, a BMW 5 Series and a Mercedes A160. The pilot flew for First Choice Airways. His two sons were, respectively, an Internet consultant and an investment banker. Both had attended university.

'It is too expensive to train. There is no job security, or even guarantee you will ever get one'. This respondent had two sons and a daughter, all in education. He flew for Dragonair, a Hong Kong-based regional and international scheduled and charter carrier. Before Dragonair he had flown for Milair out of the Hamilton Islands and Monarch out of London Gatwick.

Another respondent recommended civil engineering, the law or medicine on grounds that they required 'wider knowledge'. This respondent also worked for Dragonair. Another recommended the natural sciences on grounds that they 'Allowed you to think more freely'. This respondent had left school before A-levels.

# Appendix 6

# Pilots' Concerns and Professional Culture

## Introduction

These data, derived from the questionnaire, document respondents' worries, from 'exploitative' managements to early retirement. In aggregate, the responses also provide an insight into pilots' professional culture (their beliefs and behaviours).

## Commercial pilots' concerns

Sixty-four percent of respondents mentioned safety amongst their concerns. Twenty-five percent mentioned job security amongst their concerns. Eight percent mentioned roster stability amongst their concerns.

Nine percent of respondents mentioned both safety and job security amongst their concerns. Seven percent mentioned both safety and roster stability amongst their concerns. One percent mentioned both roster stability and job security amongst their concerns. Regarding safety, respondents made a range of comments, a selection of which are listed here.

> 'Perhaps (with hesitation) re-regulation of the industry [may address safety issues]'.
>
> 'Change duty/rest requirements and strictly enforce them. Schedules should be checked not only for legality but for efficiency and safety issues ... common sense should prevail'.
>
> '[There should be] realistic Flight Time Limits (FTL) legislation and a more involved Joint Aviation Authorities review of ATC standards in Greece and Turkey'. This respondent had been critical of 'European ATC'.
>
> 'There should be better training, better CRM training and better FTL schemes. British Airways should become the market leader, beating all opposition'. This BA pilot of three years standing had remarked upon 'Other airlines' poor training and FTL schemes which could affect our flights – altitude busts, mid-airs, fatal errors due to fatigue, etc'.
>
> 'Low morale should be improved by better man-management throughout the industry'.
>
> 'There should be better regulation and less national interest'. This respondent had bemoaned 'The fact that we do not yet have an international language for pilots'.

'Passengers should learn that this is a very expensive and sophisticated form of travel – and should be worried about the safety aspect of tickets being too cheap. Sensible fares would reduce "short-termism"'.

'We need regulation by government not voluntary self-regulation, because if there is a loophole it WILL be exploited'.

'It is important that company directors have knowledge of the work carried out by their staff – preferably, but not necessarily, they should have piloting experience'. This respondent had commented: 'Aviation operators are too much governed by accountants who do not appreciate the work undertaken by flight crew with the result that, on occasion, training and also lifestyle suffers'.

'Airlines should employ pilots with 2,000 to 3,000 hours experience ahead of low-hours college-qualified types'. This respondent had bemoaned 'Pilots joining airlines with no general aviation experience, i.e. not serving an apprenticeship and expecting command with only limited experience'.

'Early retirement'. This respondent had commented on the stress induced by contemporary rostering practices. He saw early retirement as part of the solution.

'Increase retirement age to sixty-five'. This respondent had cited 'early retirement' as one of his worries.

'Let the power reside where the blame will lie – in the cockpit'.

'Caution with the budget airline model. EVERY aircraft must have a final manual reversion capability ...'. This 74-year-old respondent had commented that ticket prices should not be cut 'at the expense of safety' and that there was 'too much automation': 'I do not like it when I see ticket prices cut to the bone and pilots having to pay for aircraft "type-ratings". The passengers must surely bear the cost of such within his ticket – not the "young man by having to add to his mortgage". This is WRONG and should be stopped at Regulatory level ...'.

'We need better quality engineering. Do not contract-out engineering. Train and use your own engineers; FTLs for long-haul in particular are not realistic; we need more time off between flights and management need to be made aware of the lifestyle of pilots and their families; management need to participate in CRM programmes with pilots, regularly'. This respondent had claimed there was 'No respect from upper management with regards quality of life'.

'Companies need to value and listen to their employees'.

'We need to prioritise funding on safety issues and crew numbers. There should be less spending on re-branding and wasteful marketing'.

'Reduced vertical separation in the cruise over Europe needs constant training and awareness, especially by ATC'.

'ICAO/EU/JAA/FAA to negotiate an ICAO scheme with fixed flight time limits. There should be mandatory sanctions against transgressors'. This respondent had bemoaned 'The lack of a world-wide (ICAO) FTL scheme' and 'Lack of safety oversight by some civil aviation authorities'.

'There should be more simulator time and corporate CRM awareness'. This respondent had highlighted a 'lack of pure flying skills enhancement' and what he called a 'Cost-above-safety culture'.

'There should be more resources for crew training and airlines' management structure should be overhauled to allow crew input'. This respondent had criticised what he called 'The management technique of rule by fear'.

'Safety should be the number one priority. We should not just pay lip service to safety. We should support the unions [on safety matters]'.

'Airlines must not be run by accountants. EVERYTHING comes down to cash'.

'There is no solution [to the management attitude that finance is everything]. It is a sad reflection on the screwed-up [confused] society in which we live. The solution would have to work its way through from primary education, so as to restore some decent moral values to society'. This respondent had criticised: 'Commercial pressure; a management attitude that finance is everything; performance-related pay awards when a human resources desk-jockey tells me how well I am flying my aeroplane, and consequently what my pay-rise will (or will not!) be. The industry has gone crazy'. This ex-RAF Flight Lieutenant flew for Royal Brunei.

'We need more respect from the management group, non-partisan Safety Departments and rules and regulations that do not contain the words: "Where possible; may; taking into account all the factors". Look at CAP 371 for a good example'. This respondent had criticised 'Safety Departments who play the blame-game (though publicly deny it) and rules and regulations where the intent is warped to fit the expedient or cheaper option'.

'All civil pilots need a few years in the military'. This ex-RAF Flight Lieutenant had commented: 'As automation increases, airmanship declines'.

'Boards should have more representatives from industry'. This respondent had criticised 'Companies run by accountants with no knowledge of the industry'.

'We need more prescriptive safety regulation; compulsory use by airlines of fatigue prevention programmes; more restrictive FTLs'.

'There is a need for standardised technology with seamless transitions between ATC agencies'. This respondent said there were problems with 'airspace control'.

'C'mon'. This pilot had criticised: 'Security; LCCs; accountants running the airline'. For this SAA pilot the solutions were self-evident.

'One must ensure that commercial considerations are separated from the operational side of the airline by strong flight operations management'.

'We should work fewer hours. There should be better rostering. There should be legal standards for FTLs to at least truck-driver limits'.

'There must be increased awareness on the part of management that their decisions have a direct impact on safety issues ...'.

'Training requirements must be increased; we must stop hiring cadets into large aircraft. Alternatively, airlines need to co-operate with second and third-tier airlines to find placements for cadets. Only the piloting profession can solve this problem. The unions at major airlines need to allow regional pilots a place in their seniority system'. This respondent had commented: 'The experience level of cadets is too low to be operating large, fast jet aircraft'.

'We must recognise that low-cost, high-intensity flying with very short turnarounds is much more fatigue-producing than conventional aviation work patterns'.

'Shareholders, Directors and managers should be less greedy'. This respondent had criticised 'The total orientation to profit-making'.

'Better FTLs; better control at smaller airports; better judgement at gates regarding drunk/drugged passengers'.

'CAA/JAA must be more proactive in making airline managements take safety seriously'.

'Governments must give regulatory agencies the power to resist the financial and industrial lobbies'. This respondent had criticised 'Duty times that induce poor performance in the cockpit'.

'We need better rostering with less hanging around; we need to pay security personnel properly and we need more research and transparency from the airline industry'.

'We need airlines to train new pilots – not just strain them. We need airlines to train new engineers – not just poach them. We need new airlines to invest more in their start-up years'.

'We need more sensible scheduling and a change in emphasis from "The Captain will normally use discretion to continue" to "The Captain will only use discretion to continue in exceptional circumstances"'.

'Schedule flight and cabin crew and aircraft together for the entire duty'.

'The industry must listen to pilots!' This respondent bemoaned 'fatigue levels and rostering practices'.

'We need to keep "cheap" labour from Eastern Europe from flying domestics in Western Europe for a $1,000 per month salary'. This respondent flew for SAS out of Oslo.

'Airlines must use FTLs as a guide – not a target'. This respondent had said that 'Duty hours, fatigue and fatigue!' were the three things that worried him most about commercial flying.

'We need stronger unions, with BALPA not just a BA mouthpiece. Sort out Ryanair as the airline is driving down standards'. This respondent flew for DHL.

'Give pilots the same professional status as doctors and lawyers'.

'Employers must consider and respect the people working for them'.

'Give more money to African nations with poor ATC to pay for education and equipment'.

'We need to get management and politicians to listen to pilots and controllers'.

'Accounts must show a little more respect towards flight crews. Training must be more thorough – I have never flown with a helicopter pilot that can land an aircraft properly. More time should be spent practising cross-wind landings on small aircraft. Finally, management must take a sensible line when dealing with reported minor offences'. This respondent flew for easyJet.

'The industry used to be run by managing professionals. Now it is run by professional managers. This is a bad thing'.

'Teach management to manage and not just promote pilot mates from your old squadron – that is, get professional human resources managers'.

'Install a management that appreciates that pilots are not idiots working on a production line or labouring on a building site. Pilots like to be informed and included in the management loop'.

'We need a new Managing Director. We need to reduce military airspace to give commercial flying more room'. This respondent flew for Ryanair. He had bemoaned 'Pressure from the MD to perform quickly and cheaply'.

'Lots of people comment on the work and lives of commercial pilots and some of them make the rules up by which we live and work. These are largely from men and women in grey suits in cosy offices with a 9–5 existence and weekends off .... I would very much like these suits to hang around on jump seats and work our rosters with us a bit more and see the effect they have on our lives at work and at home. There is an awful lot that is wrong but the CAA turns a blind eye to it. The union has an uphill struggle battling against companies that see pilots merely as large costs associated with costly aircraft and not as people who try to do a skilled job in a difficult environment. Pilots are treated in commercial aviation in a way that would not happen or be tolerated in other professions. Pilots are getting weary and wondering how they ended up in an industry that treats them so badly when all they wanted was to do something they once loved so much. We only want to be treated as human beings. Is that asking too much?' This respondent said of one of his flying jobs: 'I spent four nights a week in a poor Amsterdam airport hotel for seven months in 1999/2000. It drove me bonkers'.

Appendix 7

# Pilots' Aspirations

## Introduction

These data, obtained from the questionnaire, describe respondents' hopes and ambitions. In doing so they contribute to our understanding of pilot-culture. The data also describe respondents' retirement plans.

## Commercial pilots' aspirations

These are a selection of responses to the question: 'In terms of your flying career, how would you like to see it develop over the next five years?'

> 'I fly Airbus A321s short haul. I want to fly Airbus A330/340s long haul'. This respondent said that he would not encourage his children to become a commercial pilot.
>
> 'Command on the 747 Classic and 400. Work in training'. This respondent said that he would not encourage his children to become a commercial pilot.
>
> 'Get a command. Get involved in training'. This respondent said that he would encourage his children to become a pilot.
>
> 'I am looking for a better quality of lifestyle, i.e. sensible time off; better financial reward; better esprit-de-corps amongst work colleagues to boost morale'. This respondent said that he would not encourage his children to become a commercial pilot.
>
> 'Command within the same airline'. This respondent said that he would not encourage his children to become a commercial pilot.
>
> 'Either long haul or get involved in training in my current airline'. This respondent said that he would not encourage his children to become a commercial pilot.
>
> 'Find a job where I could fly until I am 65 [this respondent was 58]'. This respondent said that he would not encourage his children to become a commercial pilot.
>
> 'Doing part-time work so I can be with my family more and be much less stressed!'. This respondent said that he would not encourage his children to become a commercial pilot.

'I'm too close to retirement for my ambitions to matter [this respondent was fifty-six]!'. This respondent said that he would not encourage his children to become a commercial pilot.

'Alas no-one will employ me any more now I'm 60 years old [this respondent had been flying for First Choice]'. He gave as his 'reason for leaving' his last job as 'Age discrimination at 60 years old'. This respondent said that he would not encourage his children to become a commercial pilot.

'I will still be sitting in the right hand seat because Compulsory Retirement Age will be increased ... fewer Captains will retire'. This 33-year-old respondent, who flew for BA out of Heathrow, said that he would encourage his children to become a commercial pilot.

'I would like to join an operation based in a "tough" place on a one month on, one month off schedule, with the appropriate consideration and salary'. This respondent flew for Air Mauritius. This respondent said that he would not encourage his children to become a commercial pilot.

'I would like a command and eventually a training command – unlikely in the next five years'. This 38-year-old respondent said that he would not encourage his offspring to become a commercial pilot. He flew for First Choice Airways.

'Get back to the left-hand seat having given it up to change companies. That's another kettle of fish – how many experienced doctors change employers and have to go to the bottom of the pile and become an apprentice again? I would like to get involved with training to develop my skills further, put something back in, help people along and get more out of increased variety and enable more varied use of my brain than just "button-pushing" an Airbus between LHR and MAN four times a day, twenty times a week, etc'. This 39-year-old said he would not encourage his offspring to go in to the profession (even though he and his live-in partner – who worked in the banking sector – owned a house valued in 2004 at £600,000).

'I would like to get into senior management'. This pilot said he would not encourage his children to become a commercial pilot. He flew for Dragonair.

'I retire in five years, so possible part-time working as I get older'.

'I want to stay employed. I would like to move on to larger aircraft with a major'. This respondent flew for BMI Regional.

'I would like to still be flying and retain a current medical in five years'. This 49-year-old pilot flew for easyJet.

'I would like my flying career to lead me into a structure that offers more lifestyle opportunities, e.g. more days off and more challenges in flying, e.g. newer destinations. This would entail ultra-long haul operations'. This 41-year-old pilot flew for Dragonair out of Hong Kong. He said he would not encourage his children to become a commercial pilot.

'Better rostering would be nice'. This respondent who flew for Monarch said he would not encourage his children to become a commercial pilot (his response was 'NO NEVER. Appalling lifestyle').

'Slowly'. This 55-year-old pilot who flew for Nippon Cargo Airlines said he would not encourage his children to become a commercial pilot.

'I want to work part-time as soon as the home repayment scheme permits'. This 46-year-old respondent flew for Austrian out of Vienna.

'I would like to move to a prominent LCC. When there I would like to upgrade as quickly as possible and start to build seniority at a company I can feel comfortable working for until retirement'. This respondent flew for SkyWest Airlines.

'I want to gain experience in the left-hand seat with easyJet. I may move to Spain with easyJet if the opportunity arises'. This 34-year-old respondent said he would encourage his children to become a commercial pilot. He had been flying for easyJet out of Glasgow for about a year. He had previously flown for British World. He was not in a relationship and had no children. Both his parents were still alive.

'Stay employed!' This 47-year-old respondent had left school with four O-levels. He had two daughters, one ten years old the other seven. He was divorced. He said he would not encourage his children to become a commercial pilot.

'I would just like to stay in work and not be made redundant AGAIN'. Since getting his first flying position in 1982 this 48-year-old pilot had been made redundant four times. Also he left one airline (Okada, flying out of Lagos) citing a 'desire to live'.

'Triple my salary and fly the latest equipment'. This respondent flew for South African Airways.

'More time off'. This respondent, who had a son and daughter, said he would not recommend they take up a flying career.

'I would like a command in the next two years'. This 53-year-old respondent flew for Air New Zealand. He said he would not encourage his children to become a commercial pilot.

'I do not intend to fly in the next five years. I might take a short-term contract for cash'. This pilot had resigned from his last flying job (with Air Atlanta Europe) to pursue a teaching career. He said he would not encourage his children to become commercial pilots.

'I would either like to join a better airline – one I could see myself working for for the rest of my life – or I would like a command and possibly a training role'. This 27-year-old pilot flew for Bmibaby. He said he would encourage his children to become a commercial pilot.

'I want to be involved with leading-edge technology. I am looking forward to the Airbus A380'. This 44-year-old pilot flew for Emirates. He said he would encourage his children to become a commercial pilot.

'If I keep my job I will eventually increase my involvement with aviation safety and security, which I am already involved with through my pilot association'. This respondent had been furloughed by Icelandair (in 2002), had gone to work for Air Atlanta, and had then been re-hired by Icelandair.

'I want to fly larger equipment for more pay'. This 50-year-old pilot flew for SkyWest. He and his working wife owned a four-bedroomed house valued in 2004 at $650,000.

'Flying for a [US] regional I would like to see larger aircraft and more stability, either by striking out on our own or merging with other regional carriers to strengthen our position. I am pretty much where I want to be – I just want to be able to count on my employment remaining constant, if that is possible in this profession'. This respondent flew for SkyWest. He said he would not encourage his children to become commercial pilots.

'I would like to leave it and move to another career'. This 42-year-old ex-RAF pilot commenced his commercial flying career with Air 2000 in 1994. In 1996 he moved to Royal Brunei, flying out of Brunei. He was divorced. Currently in a relationship he said he would not recommend commercial aviation as a career for his offspring. He believed that rostering played a part in the breakdown of his marriage. As he explained: '[I was] not there to solve problems. Too much responsibility devolved to wife and too much temptation for both of us'.

'I just want to keep a job!' This 50-year old ex-hydrographic surveyor commenced his commercial flying career in 1980. During his career he had flown for five companies. He had been made redundant once. He had not been out of work for more than a year.

'Long-haul skipper, then training'. This 42-year-old ex-RAF pilot had flown for First Choice Airways for eight years.

'I wish to retire in one year at age 55. A part-time job on half salary would be attractive'. This 54-year-old pilot had flown for Cathay Pacific out of Hong Kong for 24 years.

'I want to find a low-stress flying job, with increasing training, simulator and check-duties'. This 57-year-old ex-RAF pilot had been made redundant from Crossair in 2003. He then worked for Jetmagic out of Cork until the company ceased operating. He said he would recommend a flying career to his offspring. 'I still think it is a worthwhile career' he said.

'I am being forced to retire soon as I fly single-pilot public transport. Am attempting to change legislation!' This 57-year-old pilot had flown for Careflight out of Stansted since 1988. He said he would recommend a flying career to his offspring (even though he felt the industry was guilty of age discrimination). As he put it: 'It's the best fun to be had and be paid for'. (His 32-year-old son worked in tourism. His 31-year-old daughter was a homemaker.)

'I would like to do less night flying and return to scheduled short-haul flying'. This 45-year-old pilot flew for Thomas Cook. He and his working wife had no children. He said that were he to become a father he would encourage his children to consider a career in commercial aviation.

'I would like an Airbus position in my present company, leading to a command on type'. This respondent flew for Air Contractors out of Dublin. In 2004 the company operated a mixed fleet of A300B4-100s and ATRs.

'I would like to improve my salary and pension benefits while maintaining my current work pattern'. This respondent worked for Scotia Helicopters. He said he would recommend a career in commercial aviation to his children.

'I aim to progress to Captain, then hopefully to make use of my training/CRM background. I am considering setting up my own CRM training business'. This 40-year-old ex-RAF pilot worked for easyJet. Originally a pharmacist he became a Squadron Leader. He said he would recommend a career in commercial aviation to his children.

'Retirement back to Canada. Buy a boat ... '. This 57-year-old ex-military pilot flew for Excel out of Manchester. He said he would recommend a career in commercial aviation to his children.

'I want a command'. This ex-RAF engineer commenced his commercial flying career with Manx in 1988. He joined BA in 1989, and was still flying with them. He said he would recommend a career in commercial aviation to his children because 'It's a passion'.

'I want a command, better rostering and reduced working hours (75% or 50%)'. This pilot, who flew for BMI, said he would recommend a career in commercial aviation to his offspring. He had worked in engineering until 1998 when he went to Oxford Air Training School. He joined BMI in 1999.

'I want to find a flight-deck seat, become a Captain and/or have management responsibilities in an airline'. This 41-year-old respondent had left his engineering post at Swissair Technics to train as flight crew. He commenced his career with Crossair in 1999, but was made redundant (with many others) in 2003.

'I want to improve my skills as a trainer while helping others'. This 34-year-old respondent flew for CHC Scotia helicopters. He said he would recommend a career in commercial aviation to his children because flying was 'good fun'.

'I want to obtain a command and change type of flying to small aircraft – police/emergency services'. This 30-year old respondent flew for CHC Scotia helicopters. He said he would recommend a career in commercial aviation to his children. 'If it is the career they want to follow I will encourage them as I enjoy it ...'.

'I would like to stay with my current employer and perhaps change fleet or move into training'. This respondent had been flying for Virgin Atlantic since 1996.

'I want to remain retired, please!' This respondent had retired from First Choice in 2000. He had commenced his flying career with Laker Airways in 1977.

'I want to settle down to 500–600 hours flying per year'. This respondent was a self-employed helicopter pilot.

'Less aircraft training; more simulator training; more time to perform this duty; less line flying'. This respondent flew for CHC Scotia.

'Possibly training, but not in the present working environment. Otherwise a slow decline into retirement in seven years'. This 52-year-old pilot flew for Bmibaby. He said he would not recommend a career in commercial aviation to his children because 'The job has become a grind. To keep up the pace of work required by today's operators for a whole career would be health-threatening'.

'I want to spend two years in a First World airline, then instruction (three-quarters time) to be with young, bright people and pass something on'. This respondent flew for Royal Brunei out of Brunei.

'I want promotion to Captain and an involvement in the training department as a line then base trainer'. This 42-year-old pilot flew for First Choice. He said he would not encourage his children to pursue a career in commercial aviation.

'I want to do more long-haul trips so I face less commuting to work'. This 52-year-old pilot flew for Britannia out of Manchester, Birmingham and Nottingham East Midlands. He lived in Egginton, Derbyshire.

'I want acceptance by accountants that pilots cannot be productive 24/7 × 12 months. I want to see pilots be accepted as professionals rather than skilled machine operators'. This respondent said he would not encourage his children to pursue a career in commercial aviation. He flew for First Choice. He attributed his divorce in part to rostering patterns. As he explained: '[Rostering played a part] by detaching me from the normal socio-economic week-based environment. Also the continual emphasis on productivity dictates that the early to late weekly roster given to UK charter pilots ensures unreal fatigue at the season's end'. He cited 'personal problems' as the reason for leaving Britannia in 1993.

'I want to gain promotion to Captain'. This 44-year-old ex-RAF Squadron Leader flew for First Choice. He said he would not recommend a flying career to his children.

'I want to get a command at DHL or move to BA'. This 35-year-old had been flying for DHL since 2004.

'I want to join a larger company with better prospects, pay and pension'. This 35-year-old ex-estate agent and office manager flew for London Executive Aviation out of Stansted.

'Happy as is. I have no wish to get involved in management or training'. This 33-year-old respondent flew for easyJet. He had attained four grade 'A' A-levels and a first class Honours degree in Aeronautical Engineering from Imperial College, London.

'If I can keep my current job I will be happy. I got into this game just in time to get on the "good" pay scale. If I make Captain in five years, that would be a pleasant surprise'. This 37-year-old father of two young children flew for SAS. He joined the company in 1998.

'I'd like to be working part-time or more regular hours'. This respondent, who had worked for BMI out of Heathrow, said he would not encourage his offspring to go in to commercial aviation.

'Captain then Training Captain'. This 45-year-old ex-RAF Squadron Leader had been flying for a UK-registered airline since 2000. He said he would not recommend a commercial flying career to his two teenage children. As he explained: 'It's not flying any more: it's managing a computer and hoping that you never actually have to use your flying skills if something goes wrong'.

'Not much. Been there, done that, and I have the T-shirt to prove it!' This 56-year-old was a career pilot. He began his flying career in 1972.

'I would simply like to find a stable airline that would give stable employment at a stable regional base'. Since beginning his career in 1979 this pilot had resigned from three positions and had been made redundant twice. He had spent two periods as a freelance. He had two teenage children. He said he would not recommend commercial flying to his children.

'Maybe a better management job in a bigger and more progressive carrier (LCC?)'. This 56-year-old ex-RAF Flight Lieutenant flew for Global Supply Systems, an international cargo carrier, out of Stansted.

'I would like to be working for a low cost carrier'. This respondent flew for SkyWest.

'I would like to work part-time until I am 60 years old, then retire completely'. This 55-year-old pilot flew for Air Atlanta Europe. He commenced flying with BA in 1970, leaving in 2004. He said he would not recommend a flying career to his children.

'I will be happy to stay in work at EMA with my present company. Anything else will be a bonus'. This 49-year-old pilot flew for BMI Regional. One of his previous employers had gone into liquidation. He said he would not recommend a flying career to his children.

'I would like to get my command (but unlikely as no/few places forecast)'. This 43-year-old ex-RAF Squadron Leader had been flying for First Choice for four years. He said he would not recommend commercial flying to his offspring.

'I would like to become a Captain/Trainer'. This 32-year-old commenced his flying career with BAC Express in 1996. He then moved to his present position with Thomas Cook in 1997.

'I would be happy to have no increase in workload, pay increases matching inflation and the company to stay solvent'. This 46-year-old ex-RAF Squadron Leader had been flying for Monarch since 1998. He said he would not recommend a flying career to his children. 'Work is too fatiguing for the rewards on offer' he stated.

'I would like a command in due course but in the mean-time become a Training First Officer and also continue my work as a Test Pilot Consultant'. This 42-year-old ex-RAF Squadron Leader flew for British Airways.

'I would like to see bigger equipment with better pay, of course. I've been working on my Masters Degree with a safety specialisation. I hope to include this in my duties in the future'. This 40-year-old ex-USAF Staff Sergeant Flight Simulator Technician flew for SkyWest. He had obtained a degree in Aeronautics from Embry-Riddle Aeronautical University in 1997, having commenced his studies in 1987. After completing his first degree he commenced a Masters Degree in Aeronautical Science at the same university. He expected to complete in 2004. He started flying for SkyWest in 1994.

'I would like to obtain my command or pursue a long-haul career and not be worked to the bone!' This 31-year-old easyJet (ex-GO) pilot said they would recommend a commercial flying career to their offspring. The respondent cohabited with a pilot (previously married). The respondent had been made redundant from BMI in 2001.

'I would like more opportunity to progress my career. That is, I would like more commands to be available'. This 36-year-old pilot used to work for BA as long-haul cabin crew. She worked as a pilot for Debonair and Sabena before joining First Choice.

'I would like to upgrade to Captain within one to two years and possibly move on to a successful, financially stable major airline. Or possibly stay here at my stable, financially conservative regional airline'. This 29-year-old pilot flew for SkyWest.

'I would like a command within 24 months and possible directing towards a training position'. This 24-year-old, ex-RAF Flying Scholarship pilot flew for GB Airways.

'I would like a better job, including a pension, better pay and a command'. This 34-year-old pilot flew for Interflight.

## Retirement plans

The following is a sample of responses to the question 'What plans, if any, do you have for your retirement?'

'I'm not sure if I am looking forward to it. We [presumably pilots] seem to pass away very quickly after retirement. Perhaps we need the stimulation'. Age 44.

'None yet!' Age 27.

'[I]nvest in property to retire before 60. Then have more time for gliding and family before keeling over'. Age 39.

'None as yet – write a book, possibly'. Age 40.

'I plan to retire at 57. I'll do part-time work, probably unrelated to flying'. Age 53.

'I have no plans. I see no reason why a healthy individual should not work until 70/75 years. I can see an "ageist" court case occurring'. Age 45.

'I aim not to retire ... I'll be forced to retire soon as I fly single-pilot public transport. Am attempting to change the legislation!' Age 57.

'Unable to retire at 60 in the way I am accustomed as we have no pension plan and salary does not keep up with the cost of living. Also my salary is paid in local currency which devalues faster than First World currencies'. This 47 year old flew for Air Mauritius.

'I aim to retire at age 38, continue flying whilst looking after my own businesses'. This 34 year old was born in Venezuela.

'Buy a sail boat and sail around the world for a year'. Age 52.

'I would like to retire at age 55, but it's now looking nearer to 60'. Age 43.

'I intend to retire as early as possible, buy a holiday home in Italy and a narrowboat'. Age 46.

'We have a smallholding which I look forward to working on'. Age 44.

'Retire to farming in Scotland with the financial security of a pension'. Age 41.

'I'll retire back to Canada, buy a boat and stack it up on gin bottles!'. Age 57.

'I want to find a simple flying job while I'm still healthy'. Age 58.

'I would like to return to flying simple aeroplanes, flying instruction at a club, perhaps, and flying for pleasure'. Age 49.

'Spend more time at home with my wife. Travel on holiday more'. Age 55.

'Sleep at night. Socialise at weekends. Plan travel beyond this month's roster'. Age 60.

'I have more hopes than plans. Having two children at university is expensive. Part-time instruction, boat and aircraft-building. Also sailing and flying my homebuilt (and spoiling my wife and children and grandchildren)'. Age 55.

'I'm saving up to buy a place in the French Pyrenees to run as a gîte'. This 33 year old Englishman flew helicopters out of Aberdeen.

'I intend to buy holiday properties for rental income'. Age 45.

'I would like to work with a church'. Age 36.

'Having been forced out of a "Final Salary" pension scheme (for a second time!) I am having to patch pension provisions together and may have to work beyond normal airline retirement ...'. This 49 year old flew for Eastern Airways out of Leeds-Bradford.

'I dread the idea. I would like to go to university'. This 56 year old had two A-levels.

'I am retired!!! But have chosen to continue flying for another operator'. This 55 year old spent 34 years at British Airways before retiring to join Air Atlanta Europe.

'I plan to retire earlier rather than later. Open a bed and breakfast!' Age 38.

'My plan is to keep working'. Age 58.

'I have no plans at present. Given the big difference in pay between First Officer and Captain, everything depends on if and when I gain promotion'. This 44 year old had joined the RAF after leaving school.

'My plans depend on whether I get the promised pension. Perhaps I'll live abroad'. Age 45.

'Travel for fun, study, read ... have a life'. Age 48.

'I have not thought much about it'. This 33 year old had been flying for seven years.

'Hopefully retire with my husband [a Training Captain] who is ten years my senior. We both have several pensions (none of which are final salary!) and a couple of investment properties. My husband hopes to retire at 60/62 years old'. Age 36.

'I hope to avoid retirement and continue working, maybe part-time. My pension will not be very good. I am considering buying a small house to let ...'. This 49 year old had flown for three UK-registered carriers. The longest time he had spent at a carrier was 17 years.

'My plan is to live long enough to achieve retirement!' Age 51.

'Fly sailplanes. Maybe live in the USA for a few years. Build small experimental aircraft'. Age 49.

'I have not thought about it much as yet. A pilot-colleague died recently after only one month of retirement ... there might not be much point in planning too far!' Age 37.

# Bibliography


Abbas, Z. (2005), 'Pakistan's First Women Fighter Pilots', *BBC News*, 11 May. <http://news.bbc.co.uk/>

*Airliner World* (2005a), 'Delta and Northwest File For Chapter 11', *Airliner World*, November.

*Airliner World* (2005b), 'FlyBe Studies UK Regional Expansion', *Airliner World*, November.

*Air Safety Week* (2004a), 'Experts: Human Fatigue Bigger Risk to Safety Than Realised', *Air Safety Week*, 22 March.

*Air Safety Week* (2004b), 'Flight Attendants Union Launches Fatigue Survey', *Air Safety Week*, 19 April.

Ambrose, S.E. (1985), *Rise to Globalism. American Foreign Policy Since 1938* (Harmondsworth: Penguin Books).

Arlidge, J. (2001), 'How low can they go?' *The Observer*, 19 August.

Ashcroft, M. (2004), 'The Ryan king', *British Industry*, September.

Baker, C. (2005), 'European carriers report mixed results', *Airline Business*, September.

Baker, C. (2006), 'BA swings its management axe', *Airline Business*, January.

Baker, C. and Field, D. (2005), 'Fare rises limit fuel pain', *Airline Business*, September.

Ball, C. (2005), 'The fruits of labour', *The Guardian*, 26 November.

Ball, R. (2003), 'Technology Troubles', *The Log*, June.

Ball, R. (2005), 'Majority view', *The Log*, June.

Barber, B. (2002), 'Unions must have a voice on the board', *The Observer*, 22 December.

Barry, S. (2005), 'Where are the women pilots? *The Log*, February.

Bartol, K.M. and Martin, D.C. (1998) *Management* (New York: Irwin/McGraw-Hill).

Beaty, D. (1969), *The Human Factor in Aircraft Accidents* (London: Secker and Warburg).

Beck, U. (1992), *Risk Society: Towards a New Modernity* (London: Sage).

Bennett, S.A. (1996), 'Getting the Public on Board'. *Chemistry in Britain*, 32:4, April, 39–41.

Bennett, S.A. (2001), *Human Error – by Design?* (Leicester: Perpetuity Press).

Bennett, S.A. (2002a), 'Lock and Load? Explaining Different Policies for Delivering Safety and Security in the Air', *International Journal of Mass Emergencies and Disasters*, 20:2, August, 141–69.

Bennett, S.A. (2002b), *Victims of Hubris? The Decline and Fall of Pan Am and Swissair* (Leicester: University of Leicester Department of Criminology).

Bennett, S.A. (2003), 'Flight crew stress and fatigue in low-cost commercial air operations – an appraisal', *International Journal of Risk Assessment and Management*, 4:2, 207–31.

Bennett, S.A. (2005), 'Aviation needs a *lingua franca*', *The Aerospace Professional*, July.

Bennett, S.A. and Shaw, A.P. (2003), 'Incidents and accidents on the ramp: does 'risk communication' provide a solution?' *Human Factors and Aerospace Safety*, 3:4, 333–52.

Bhargava, K. (2005), 'Budget Carriers Crowd Over India', *Airliner World*, November.

Bijker, W.E. and Law, J. (1992), 'General Introduction', in W.E. Bijker and J. Law (eds), *Shaping Technology/Building Society. Studies in Sociotechnical Change* (Cambridge: MIT Press).

Billings, C.E. (1997), *Aviation Automation: The Search for a Human-Centred Approach* (New Jersey: Lawrence Erlbaum Associates).

Black, I. and Wintour, P. (2001), 'UK caves in on work directive', *The Guardian*, 12 June.

Blogging at FL250 (2005), 'The Life of an Airline Pilot. Monday, February 28, 2005' <http://fl250.blogspot.com/2005/02/life-of-airline-pilot.html/>

Bogdanor, V. (ed.) (2005) *Joined-Up Government* (Oxford: OUP/British Academy).

Bolton Museums (2005), 'Mass-Observation' <http://www.boltonmuseums.org.uk/HTML/spender/history_mass_observation.html/>

Bonner, A. (2003), *Media Relations* (Wasaga Beach: Briston House).

Bor, R. (ed.) (2003), *Passenger Behaviour* (Aldershot: Ashgate).

Bourke, J. (2005), *Fear: A Cultural History* (London: Virago).

Bridgewater, S. (2003), 'So You Want to be a Pilot?', in M. Nicholls (ed.), *Century of Flight* (Stamford: Key Publishing Ltd).

British Air Line Pilots' Association (2004), 'Sub part Q on Flight Duty Time' <http://www.balpa.org/intranet/BALPA-Camp/FLTs/CEO-warning.doc_cvt.htm/>

British Air Line Pilots' Association (2005a), *How to become a Commercial Pilot* (Staines: BALPA) <http://www.balpa.org/>

British Air Line Pilots' Association (2005b), *News from BALPA. January 10 2005. Woman Pilot Takes BA to Employment Tribunal on Discrimination* <http://www.balpa.org/>

British Airways (2005), *Explore Our Working World* <http://www.britishairwaysjobs.com/>

British Broadcasting Corporation News (2005), 'BA pilot wins discrimination case' <http://news.bbc.co.uk/go/pr/fr/-/1/hi/england/dorset/4471851.stm/> 22 April.

Buck, R.N. (1994), *The Pilot's Burden: Flight Safety and the Roots of Pilot Error* (Ames: Iowa State University Press).

Bunting, M. (2004), 'Work-life balance', *The Guardian*, 24 September.

Cabinet Office – Prime Minister's Strategy Unit (2004), *Alcohol Harm Reduction Strategy for England* (London: Cabinet Office).

Callaghan, D. (2005), 'Reduced doctors' hours "working well", report says', *The Guardian*, 17 August.

Carayol, R. (2004), 'A bruiser in the f****** boardroom', *The Observer*, 7 November.

Carvel, J. (2003), '7-year high for divorce rates', *The Guardian*, 29 August.

Centerprise Trust (1977), *A People's Autobiography of Hackney* (London: Centerprise Trust Ltd).

Charlwood, D.E. (1968), *Take-off to Touchdown* (London: Angus & Robertson Ltd).

Childs, D. (1995), *Britain since 1939. Progress and Decline* (London: Macmillan).

Chingosho, E. (2005), 'Africa's era of liberalisation', *Airline Business*, September.

Cineflix-Galaxie-NextFilm (2004), *Aircrash Investigations: A Wounded Bird.* Broadcast on UK Channel Five, 22 December, 2005.

Clark, A. (2004a), 'Singapore in long-haul record', *The Guardian*, 3 February <http://www.guardian.co.uk/>

Clark, A. (2004b), 'Budget airlines doomed, says Ryanair chief', *The Guardian*, 2 June.

Clark, A. (2004c), 'Pilots defeat plan to impose longer working hours', *The Guardian*, 15 June.

Cottell, C. (2004), 'Unions look to get off the ground at Ryanair', *The Guardian*, 4 September <http://www.guardian.co.uk/>

Creaton, S. (2005), *Ryanair. How a Small Irish Airline Conquered Europe* (London: Aurum).

Cuff, E.C., Sharrock, W.W. and Francis, D.W. (1992), *Perspectives in Sociology* (London: Routledge).

Curtis, P. (2005), 'Report reveals economic divide among students', *The Guardian, Higher Education Supplement*, 23 November.

Davies, C. (2005), 'Woman pilot scores second victory over part-time working', *The Daily Telegraph*, 22 July <http://www.telegraph.co.uk/>

Dekker, S. (2004), 'On the other side of promise: what should we automate today?', in D. Harris (ed.), *Human Factors for Civil Flight Deck Design* (Aldershot: Ashgate).

Dienel, H.L. and Lyth, P. (1998), *Flying the Flag* (London: Macmillan).

Di Maio, A. (2004), 'Move "Joined-Up Government" From Theory to Reality' <http://www.gartner.com/DisplayDocument?doc_cd=123844/> 20 October.

Duke, N. and Lanchbery, E. (eds) (1964), *The Crowded Sky* (London: Corgi Books).

Eglin, R. and Ritchie, B. (1980), *Fly Me, I'm Freddie!* (London: Weidenfeld & Nicholson).

Elliott, L. (2004), 'Oil price soars to new high', *The Guardian*, 2 June.

Endres, G. (2006), 'Shifting sands', *Airline Business*, January.

Everard, C. (2006), 'Is safety an issue?' *Aerospace International*, February.

Eyre, E.C. (1984), *Mastering Basic Management* (Basingstoke: Macmillan).

Feddy, K. and Webster, N. (2000), 'Terror-flight passengers hail Captain Fantastic',

*Manchester Evening News*, 17 May.

Field, D. (2005a), 'New wave of US hopefuls mix luxury with low fares', *Airline Business*, August.

Field, D. (2005b), 'Wage cuts hit the wall', *Airline Business*, August.

Field, D. (2005c), 'Day of reckoning nears for ailing Independence Air', *Airline Business*, December.

Field, D. (2006), 'High stakes', *Airline Business*, January.

Fleck, F., Parsley, D. and Smith, D. (2001), 'Is Swissair the First of Many to go Bust?' *The Sunday Times*, 7 October.

*Flight International* (2005), 'EADS puts opportunities for women high on agenda', *Flight International*, 28 June – 4 July.

Francis, G. (2004), 'European Pilots Win Battle over Flight Time Limitations', *Air Line Pilot*, September.

Gandt, R. (1999), *Skygods. The Fall of Pan Am* (McLean, Virginia: Paladwr Press).

Gant, K. (2003), 'The Ninety-Nines. Women in Aviation' <http://www.ninety-nines.org/wia.html/>

Gaskell, K. (1999), *British Airways. Its History, Aircraft and Liveries* (Shrewsbury: Airlife).

Gilbert, M. (1999), *Challenge to Civilisation. A History of the Twentieth Century, 1952–1999* (London: Harper Collins).

Gilbert, N. (1993), *Researching Social Life* (Thousand Oaks, United States of America: Sage).

Greater Manchester Low Pay Unit (1995), *Workers' Voices: Accounts of Working Life in Britain in the Nineties* (Manchester: GMLPU).

Greaves, R. (2004), 'Machine Minder 1', *The Log*, December.

Hall, S. and Jacques, M. (1989), *New Times. The Changing Face of Politics in the 1990s* (London: Lawrence and Wishart).

Hancock, C. (2004a), 'Ryanair pilots to fight O'Leary ban on unions', *Times Online*, 8 February <http://www.timesonline.co.uk/>

Hancock, C. (2004b), 'Ryanair pilots seething over contract ultimatum', *Times Online*, 7 November <http://www.timesonline.co.uk/>

Hannon, B. and Kingsley-Jones, M. (2000), 'SIA 747 crashed attempting take-off from closed runway', *Flight International*, 7–13 November, p. 4.

Harkin, P. and Hunt, D. (2005), 'Terminal breakdown', *Survive Magazine*, Fourth Quarter.

Harris, D. (2005), 'Drinking and flying: causes, effects and the development of effective countermeasures', in D. Harris and H.C. Muir (eds), *Contemporary Issues in Human Factors and Aviation Safety* (Aldershot: Ashgate).

Harvey, L. and MacDonald, M. (1993), *Doing Sociology* (London: Macmillan).

Health and Safety Executive (2000), *Aircraft Turnaround* (Sudbury: HSE Books).

Heath, C.C., Hindmarsh, J. and Luff, P. (1999), 'Interaction in Isolation: The Dislocated World of the London Underground Train Driver', *Sociology*, 33:3, 555–75.

Helmreich, R.L. and Merritt, A.C. (1998), *Culture at Work in Aviation and Medicine.*

*National, Organisational and Professional Influences* (Aldershot: Ashgate).
Hennessy, P. (1993), *Never Again. Britain 1945-1951* (London: Vintage).
Hensser, H. (1953), *Comet Highway* (UK: John Murray (Publishers) Ltd).
Hewitt, P. (2001), 'Hearing other voices', *The Guardian*, 5 July.
Holloway, S. (2002), *Airlines: Managing to Make Money* (Aldershot: Ashgate).
Hudson, K. (1972), *Air Travel. A Social History* (Bath: Adams & Dart).
Hudson, K. and Pettifer, J. (1979), *Diamonds in the Sky. A Social History of Air Travel* (London: The Bodley Head Ltd. and the British Broadcasting Corporation (BBC)).
Hudson, R. and Williams, A.M. (1989), *Divided Britain* (London: Belhaven Press).
Hutter, B. and Power, M. (2005), 'Risk & Deregulation', *Risk & Regulation*, 9, Summer.
International Civil Aviation Organisation (2003), *100 Years of Civil Aviation* (Montreal: ICAO).
International Transport Workers' Federation (2004), 'Is the flight time I am on duty really okay?' *Ryan be Fair. Giving Ryanair Employees a Collective Voice* <http://www.ryan-be-fair.org/>
Ionides, N. (2005), 'India frees more markets ... and bursts with low-cost life', *Airline Business*, July.
Ionides, N. (2006) 'Dreams or Reality?' *Airline Business*, May.
Irwin, A. (1995), *Citizen Science* (London: Routledge).
Johnson, R. (2004), 'Cover story: Lord of the wings', *Times Online*, 1 August <http://www.timesonline.co.uk/>
Jones, K. (2004), 'Machine Minder 2', *The Log*, December.
Kane, F. (2005), 'Walsh passes his flight test', *The Observer*, 20 November.
Kern, T. (1997), *Redefining Airmanship* (New York: McGraw-Hill).
Kohn, R. (2005), *So You Want To Be A Pilot 2006* (London: Royal Aeronautical Society & Guild of Air Pilots and Navigators).
Koonce, J.M. (1999), 'A Historical Overview of Human Factors in Aviation', in D.J. Garland, J.A. Wise and V.D. Hopkin (eds), *Handbook of Aviation Human Factors* (New Jersey: Lawrence Erlbaum).
Krisbergs-Culp, M. (2001), 'What it's like to be a corporate pilot I' <http://www.ninety-nines.org/>
Learmount, D. (2005), 'UK drives to get greener', *Flight International*, 28 June – 4 July.
Lewis, J. (1984), *Women in England 1870–1950* (London: HarvesterWheatsheaf).
Macdonald, K. and Tipton, C. (1993), 'Using Documents', in N. Gilbert (ed.), *Researching Social Life* (London: Sage).
Maidment, R. and McGrew, A. (1991), *The American Political Process* (London: Sage).
Maslen, R. (2003), 'Air Travel. The booming years', in M. Nicholls (ed.), *Century of Flight* (Stamford: Key Publishing Ltd).
Maslen, R. (2006), 'Jazeera Airways: expanding the low-cost formula', *Airliner World*, February.

Mass-Observation Archive (2001), 'A Brief History of Mass-Observation' <http://www.massobs.org.uk/history.html/>

Matthews, A. (2005), 'Morale Boost', *The Log*, February.

McGurk, J. (2004), 'Empowered Flight in 2004', *The Log*, February.

McLaurin, H. (2005), 'The Pilot. Could you handle having endless blue skies as your daily office view? One woman wouldn't swap it for the world', *ivillage* <http://ivillage.co.uk/>

Milde, M. (1977), 'Aviation Law', in D. Mondey (ed.), *The International Encyclopaedia of Aviation* (London: Octopus).

Mintier, T. (1999), 'Female pilots earning respect in Gulf combat roles', *CNN*, 9 January <http://www.cnn.com/>

Morgan, K.O. (1993) *The Oxford Illustrated History of Britain* (London, BCA, by arrangement with Oxford University Press).

Morrison, S.A. and Winston, C. (1995), *The Evolution of the Airline Industry* (Washington: The Brookings Institution).

Murray, R. (1989), 'Fordism and Post-Fordism', in S. Hall and M. Jacques (eds), *New Times. The Changing Face of Politics in the 1990s* (London: Lawrence and Wishart).

Newton, L. and Ridley, N. (2001), 'Air Crash Hero Saved A Village. Bravery of Helicopter Pilot in Accident that Claimed Three Lives', *Leicester Mercury*, 1 August.

Nicholls, M. (2006), 'Out of the Blue', *Airliner World*, February.

Orlady, H.W. and Orlady, L.M. (1999), *Human Factors in Multi-Crew Flight Operations* (Aldershot: Ashgate).

Osborne, A. (2005), 'Ryanair boss blasts compensation', *The Daily Telegraph*, 5 October.

Osborne, H. (2005a), 'TUC attacks long working hours', *The Guardian*, 9 June.

Osborne, H. (2005b), 'Women earn 17.2% less than men', *The Guardian*, 10 November.

Palmason, S.T. and Kramer, J. (1999), 'How to Benefit from Real Partnership – The Southwest Airlines' Experience', *The Log*, August.

Parker, N., Thompson, P., Lea, M., Pharo, C., Troup, J. and Lazzeri, A. (1999), 'We Told Him Not to Land', *The Sun*, 16 September.

Perrow, C. (1984), *Normal Accidents* (New York: Basic Books).

Petzinger, T. (1995), *Hard Landing. How the Epic Contest for Power and Profits Plunged the Airlines into Chaos* (London: Aurum).

Pirihi, L. (2005), 'Pilots – what are they earning?' <http://www.black-personals.com/>

Pomerantz, G.M. (2001), *Nine Minutes, Twenty Seconds. The Tragedy and Triumph of ASA Flight 529* (London: Michael Joseph).

Prince, M. (1990), *Crash Course: The World of Air Safety* (London: Grafton Books).

Prospects.ac.uk (2005), 'Airline pilot in close-up' <http://www.prospects.ac.uk/>

Quinn, T. (2003), *Wings Over the World. Tales from the Golden Age of Air Travel*

(London: Aurum Press).
Read, B. (2005), 'State of flux', *Aerospace International*, June.
Reason, J. (1995), *Human Error* (Cambridge: Cambridge University Press).
Reece, M (2004), 'Taking the plunge', *The Log*, December.
Reed, A. (1990), *Airline: The Inside Story of British Airways* (London: BBC Books).
Reed, P. (2002), *Railtracks in the Sky* (Nottingham: Spokesman).
Reid, T. (2005), 'Crash survivors tell of 'miracle' escape', *The Times*, 4 August.
Rendall, D. (1996), *Jane's Aircraft Recognition Guide* (Glasgow: Harper-Collins).
Rendall, I. (1988), *Reaching for the Skies* (London: BBC Books).
Reuters (2005), 'Report: Air France crash plane landed long', Reproduced at <http://www.airdisaster.com/news/0805/05/news.shtml/>
Ryan, D. (1995), 'Asserting US Power', in P.J. Davies (ed.), *An American Quarter Century. US Politics from Vietnam to Clinton* (Manchester: Manchester University Press).
Saettone, M. (2001), 'What it's like to be a Corporate Pilot II' <http://www.ninety-nines.org/>
Sampson, A. (1984), *Empires of the Sky. The Politics, Contests and Cartels of World Airlines* (London: Hodder and Stoughton).
Scott, L. (2005), 'The sky's the limit', *Metro*, 2 September.
Scott, M. (2004), 'From airline pilot to airline machine-minder', *The Log*, October.
Shearman, P. (1992), *Air Transport. Strategic Issues in Planning and Development* (London: Pitman).
Sheppard, I. (2005), 'UK Flight Safety Committee Annual Seminar 2005. Aviation Safety: Looking Ahead 20 Years', *Focus on Commercial Aviation Safety*, 61, Winter.
*Skytrax Research* (2005), 'Singapore Airlines unveils the newest aircraft type to join its fleet – the ultra long-range Airbus 340-500' <http://www.airlinequality.com/Product/SIA-Ex-Y.htm/>
Smither, R.D. (1988), *The Psychology of Work and Human Performance* (New York: Harper & Row).
South West Public Health Observatory (2005), 'Trends in life expectancy by social class 1972-1999' <http://www.swpho.org.uk/onslifeexpect.htm/>
Stallwood, O. (2005), 'We thought we'd die on smoke-filled plane', *Metro*, 5 August.
Stallwood, O. (2006), 'Jamie "sends the cost of a school dinner soaring"', *Metro*, 5 January.
Stanton, N.A., Salmon, P.M., Walker, G.H., Baber, C. and Jenkins, D.P. (2005) *Human Factors Methods – A Practical Guide for Engineering and Design* (Aldershot: Ashgate).
Steele, J. (2005), 'A new kind of Europe', *The Guardian*, 15 June.
Stephenson, A. (2005), 'Where are the women pilots?' *The Log*, April.
Stewart-Smith, J. (2005), 'Buyer Beware', *The Log*, February.
Stone, R.B. and Babcock, G.L. (1988), 'Airline Pilots' Perspective', in E.L. Wiener

and D.C. Nagel (eds), *Human Factors in Aviation* (San Diego: Academic Press).

Sullivan, B. and Edman, J. (2004), *"Flying Partners" – Evidence for a Global Supraculture Composed of People who Fly for a Living*. Paper presented at the 33rd Annual Meeting of the Society for Cross-Cultural Research, San Jose, California, USA, 19 February.

Summerfield, C. and Gill, B. (eds) (2005), *Social Trends. No. 35. 2005 Edition* (Basingstoke: Palgrave Macmillan).

Tarry, C. (2005), 'Asia comes into focus'. *Airline Business*, December.

Taylor, J.W.R. (1967), *Aircraft, Aircraft* (London: Paul Hamlyn).

Taylor, P., Richardson, J., Yeo, A., Marsh, I., Trobe, K. and Pilkington, A. (1995), *Sociology in Focus* (Ormskirk: Causeway Press Ltd).

*The Guardian* (2005a), 'Corporate sexism is still rife', *The Guardian*, 5 August.

*The Guardian* (2005b), 'Healthy, wealthy – and still at work', *The Guardian*, 27 December.

Theobald, R. (1994), *Understanding Industrial Society. A Sociological Guide* (London: Macmillan).

Toft, B. and Reynolds, S. (1997), *Learning from Disasters* (Leicester: Perpetuity Press).

Triggle, N. (2005), 'Are GPs worth £100,000?' <http://news.bbc.co.uk/go/pr/fr/-/1/hi/health/>

Walker, C. (2003), 'Women in Aviation', in M. Nicholls (ed.), *Century of Flight* (Stamford: Key Publishing Ltd).

Walsh, G., Ungoed-Thomas, J. and Calvert, J. (2006), 'Watchdog slams BA's air safety', *The Sunday Times*, 1 January.

Walters, J. (2002), 'Happy landings? Low-cost airlines have been accused of pushing pilots to work worryingly long hours', *The Observer*, 23 June.

Ward, L. (2005), 'Progress on work-life balance is a myth for many, says report', *The Guardian*, 17 November.

Webster, B. (2002), 'Budget airline pilots accused of putting safety at risk', *The Times*, 18 June.

Wegrich, K. (2005), 'Stereotypes of (de)regulation: How easy answers to complex problems create poor policies', *Risk & Regulation*, Number 9, Summer.

Whelan, E.A. (2003), 'Editorial. Epidemiology. Cancer incidence in airline cabin crew'. *Occupational and Environmental Medicine*, 60, 805–806.

White, S. (2002) 'Air Girl's Fire Alert Over Dry Ice Fumes', *Daily Mirror*, 26 June, p. 24.

Whiteley, P. (2002), 'Stay-at-home citizens', *The Guardian*, 1 May.

Williams, D. (1997), 'Inches From Catastrophe', *Daily Mail*, 6 November.

Williams, G. (2002), *Airline Competition: Deregulation's Mixed Legacy* (Aldershot: Ashgate).

Wolfe, T. (1991), *The Right Stuff* (London: Picador).

Wood, M. (2005), 'Bloke Bashing', *The Log*, April.

Wood, S. (2004), *CAA Paper 2004/10. Flight Crew Reliance on Automation* (Gatwick: Civil Aviation Authority Research Management Department – Safety

Regulation Group).

Woodall, P. (2005), 'Fragile foundations', *The Economist: The World in 2006. 20th edition.*

# Index